Collins

Weather
ALMANAC
A GUIDE TO
2022

Storm Dunlop

Published by Collins
An imprint of HarperCollins Publishers
Westerhill Road
Bishopbriggs
Glasgow G64 2QT
www.harpercollins.co.uk

© HarperCollins Publishers 2021
Text and illustrations © Storm Dunlop and Wil Tirion
Cover illustrations © Julia Murray
Images and illustrations see acknowledgements page 258

Collins ® is a registered trademark of HarperCollins Publishers Ltd
All rights reserved. No part of this publication may be reproduced, stored
in a retrieval system, or transmitted, in any form or by any means, electronic,
mechanical, photocopying, recording or otherwise without the prior written
permission of the publisher and copyright owners.

The contents of this publication are believed correct at the time of printing.
Nevertheless the publisher can accept no responsibility for errors or omissions,
changes in the detail given or for any expense or loss thereby caused.

HarperCollins does not warrant that any website mentioned in this title will
be provided uninterrupted, that any website will be error free, that defects will
be corrected, or that the website or the server that makes it available are free
of viruses or bugs. For full terms and conditions please refer to the site
terms provided on the website.

A catalogue record for this book is available from the British Library

ISBN 978-0-00-846989-4

10 9 8 7 6 5 4 3 2 1

Printed and bound by CPI Group (UK) Ltd,
Croydon CR0 4YY

If you would like to comment on any aspect of this book,
please contact us at the above address or online.
e-mail: collins.reference@harpercollins.co.uk

MIX
Paper from
responsible sources
FSC™ C007454

This book is produced from independently certified
FSC™ paper to ensure responsible forest management.

For more information visit: www.harpercollins.co.uk/green

Contents

Introduction

Our variable weather

Anyone living in the British Isles hardly needs to be told that the weather is extremely variable. Yet it seems that whenever there is an extreme event, the media and, in particular, politicians describe it as 'unprecedented'. But it is not. Britain has always experienced extreme windstorms, snowfall, rainfall, thunderstorms, flooding and such major events. And always will. Such events may be unusual, and 'not within living memory of the oldest inhabitant', but, overall, the weather always exhibits such extremes. We may have learned from events such as the east coast floods of 1953, and constructed the Thames Barrier and the barrier on the River Hull, but a North Sea surge will occur again. The Somerset Levels have been flooded many times in the past, so the flooding in 2012 and 2014 was not that extraordinary.

The weather in Britain – its climate – is basically a maritime climate, determined by the proximity to the Atlantic Ocean. It is largely determined by the changes resulting from incursions of dry continental air from the Eurasian landmass to the east, contrasting with the prevailing moist maritime air from the Atlantic Ocean to the west. The general mildness of the climate, in comparison with other locations at a similar latitude, has often been ascribed to 'the Gulf Stream'. In fact, the Gulf Stream exists only on the western side of the Atlantic, along the east coast of North America, and the warm current off the coast of Britain is correctly known as the North Atlantic Drift. In reality, it is not solely the warmth of the oceanic waters that creates the mild climate.

The Rocky Mountains in North America impede the westerly flow of air and create a series of north/south waves that propagate eastwards (and actually right around the world). These waves cause north/south oscillations of the jet stream (the Polar Front jet stream) that, in turn, control the progression of the low-pressure areas (depressions) that travel from west to east and create most of the changeable weather. The jet stream 'steers' the depressions, sometimes to the north of the British

Isles and sometimes far to the south. It may also assume a strong flow in longitude (known as a 'meridional' flow) leading to a blocking situation, where depressions cannot move eastwards and may come to a halt or be forced to travel far to the north or south. Such a block (especially in winter) may draw frigid air directly from the Arctic Ocean or continental air from the east (usually described as 'from Siberia').

The location of the jet stream itself is governed by something known to meteorologists as the North Atlantic Oscillation (NAO). Basically, this may be thought of as the distribution of pressure between the Azores High over the central Atlantic and the semi-permanent Icelandic Low. The average route of the jet stream is to the north of the British Isles, so the whole country is subject to the strong westerlies. Northern regions tend to be mainly subject to the depressions arriving from the Atlantic and the weather that accompanies them. So overall, the climate of the British Isles may be described as: warmer and drier in the south and east, wetter in the west and north.

When the NAO has a 'positive' index, with high pressure and warm air in the south compared with low pressure and low temperatures in the north, depressions are steered north of the British Isles. Although most of the country experiences windy and wet weather, the west and north tends to be affected most, with southern and eastern England warmer and drier. When the NAO has a 'negative' index, the Azores High is displaced towards the north and the jet stream tends to show strong meanders towards the south. Frequently there is a slowly moving low pressure area over the near (north-western) Continent or over the north-eastern Atlantic. The jet stream wraps around this, before turning back to the north. (Blocks are, however, most frequent in the spring.)

Climatic regions of the British Isles
It is appropriate to deal with the climates of the British Isles by discussing eight separate regions, which are:

Sunrise / Sunset & Moonrise / Moonset
For four dates within each month, tables show the times at which the Sun and Moon rise and set at the four capital cities in the United Kingdom: Belfast, Cardiff, Edinburgh and London. For Edinburgh and London, the locations of specific observatories are used, but not for Belfast and Cardiff, where more general locations are employed. Although the exact time of Sunrise/set and Moonrise/set at any observer's location depends on their exact position, including their latitude and longitude and height above sea level, the times shown will give a useful indication of the timing of the events. However, calculation of rising and setting times is complicated and strongly depends on the location of the observer. Note that all the times are calculated astronomically, using what is known as Universal Time (UT). This is identical to Greenwich Mean Time (GMT). The times given do not take account of British Summer Time (BST). Some effects are quite large. For example, at the summer solstice in 2022, sunrise is some 48 minutes later at Edinburgh than it is at Lerwick in the Shetlands. Sunset is just 31 minutes earlier on the same date.

Edinburgh & Lerwick: Comparison of sunrise & sunset times
A comparison of the timings of sunrise and sunset at Lerwick (latitude 60.16°N) and Edinburgh (latitude 55.9°N) in the following table gives an idea of how the times change with latitude. The timings are those at the equinoxes (March 20 and September 23 in 2022) and the solstices (June 21 and December 21 in 2022). It is notable, of course, that although the times are different, the azimuth of sunrise is identical at both equinoxes, because the Sun is then crossing the celestial equator. At the solstices, both times and azimuths are different. Azimuth is measured in degrees from north, through east, south and west, and then back to north. Diagrams of twilight for both Edinburgh and Lerwick are shown on page 252.

Location	Date	Rise	Azimuth	Set	Azimuth
Edinburgh	20 Mar 2022 (Sun)	06:15	89	18:26	271
Lerwick	20 Mar 2022 (Sun)	06:07	89	18:19	271
Edinburgh	21 Jun 2022 (Tue)	03:27	43	21:03	317
Lerwick	21 Jun 2022 (Tue)	02:39	34	21:34	326
Edinburgh	23 Sep 2022 (Fri)	06:00	89	18:09	271
Lerwick	23 Sep 2022 (Fri)	05:51	89	18:02	271
Edinburgh	21 Dec 2022 (Wed)	08:42	134	15:40	226
Lerwick	21 Dec 2022 (Wed)	09:08	141	14:57	219

Apart from the times, this table (and the monthly tables) also shows the azimuth of each event, which indicates where the body concerned rises or sets. These azimuths are given in degrees, and the table given here shows the azimuths for various compass points in the eastern and western sectors of the horizon.

Table of azimuths

Degrees	Compass point
Eastern horizon	
45°	NE
67° 30'	ENE
90°	E
112° 30'	ESE
135°	SE
157° 30'	SSE
Western horizon	
202° 30'	SSW
225°	SW
247° 30'	WSW
70°	W
292° 30'	WNW
315°	NW
337° 30'	NNW

The actual latitude and longitude used in the calculations are shown in the following table. It will be seen that the altitudes of the observatories in Edinburgh (Royal Observatory Edinburgh, ROE) and London (Mill Hill Observatory) are quite considerable (that for ROE is particularly large) and these altitudes will affect the rising and setting times, which are calculated to apply to observers closer to sea level. (Generally, close to the observatories, such rising times will be slightly later, and setting times slightly earlier than those shown.)

Latitude and longitude of UK capital cities

City	Longitude	Latitude	Altitude
Belfast	5°56'00.0" W	54°36'00.0" N	3 m
Cardiff	3°11'00.0" W	51°30'00.0" N	3 m
Edinburgh (ROE)	3°11'00.0" W	55°55'30.0" N	146 m
London (Mill Hill)	0°14'24.0" W	51°36'48.0" N	81 m

Twilight

For each individual month, we give details of sunrise and sunset times (and moonrise and moonset times), together with the azimuths (which give an idea of where the rising or setting takes place – see previous page) for the four capital cities of the regions of the United Kingdom. But twilight also varies considerably from place to place, so the monthly diagrams here show the duration of twilight at those four cities. During the summer, twilight may persist throughout the night. This applies everywhere in the United Kingdom, so two additional yearly twilight diagrams are included (on pages 252–253): one for St Mary's in the Scilly Isles, and one for Lerwick in the Shetlands. Although the hours of complete darkness increase as one moves towards the equator, it will be seen that at neither London nor St Mary's is there full darkness at midsummer.

There are three recognised stages of twilight: *civil twilight*, when the centre of the Sun is less than 6° below the horizon; *nautical twilight*, when the Sun is between 6° and 12° below the horizon; and *astronomical twilight*, when the Sun is between 12° and 18° below the horizon. *Full darkness* occurs only when the Sun is more than 18° below the horizon. The time at which civil twilight begins is sometimes known in the UK as 'lighting-up time'. During nautical twilight, the very brightest stars only are visible. (These are the stars that were used for navigation, hence the name for this stage.) During astronomical twilight, the faintest stars visible to the naked eye may be seen directly overhead, but are lost at lower altitudes. They become visible only once it is fully dark. The diagrams show the duration of twilight at the various cities. Of the locations shown, during the summer months there is full darkness at most of the cities, but it never occurs during the summer at the latitude of London.

The diagrams also show the times of New and Full Moon (black and white symbols, respectively). As may be seen, at most locations during the year, roughly half of New and Full Moon phases may come during daylight. For this reason, the exact phase may be invisible at one location, but be clearly seen elsewhere in the world. The exact times of the events are given in the diagrams for each individual month.

Twilight diagrams for each of the four capital cities are shown every month, and full yearly diagrams are shown on pages 251–253.

Also shown each month is the phase of the Moon for every day, together with the age of the Moon, which is counted from New Moon.

The seasons

By convention, the year has always been divided into four seasons: spring, summer, autumn and winter. In the late eighteenth century, an early German meteorological society, the Societas Meteorologica Palatina, active in the Rhineland, defined the seasons as each consisting of three whole months, beginning before the equinoxes and solstices. So spring consisted of the months of March, April and May; summer of June, July and August; autumn of September, October and November, and winter of December, January and February. There has been

---

Föhn effect

When humid air is forced to rise over high ground, it normally deposits some precipitation in the form of rain or snow. When the air descends on the far side of the hills, because it has lost some of its moisture, it warms at a greater rate than it cooled on its ascent. This 'föhn effect' may cause temperatures on the leeward side of hills or mountains to be much warmer than locations at a corresponding altitude on the windward side.

---

a tendency by meteorologists to follow this convention to this day, with winter regarded as the three calendar months with the lowest temperatures in the northern hemisphere (December, January and February) and summer those with the warmest (June, July and August). Astronomers, however, regard the seasons as lasting three months, but centred on the dates of the equinoxes and solstices (March 20, September 23 and June 21, December 21 in 2022).

Some ecologists tend to regard the year as divided into six seasons. Analysis of the prevailing weather types in Britain, however, suggests that there are five distinct seasons. (The characteristic weather of each is described in the month in which the season begins.) Although, obviously, the seasons cannot be specified as starting and ending on specific calendar dates, it is useful to identify them in this way. So in Britain, we have:

Early winter
> November 20 to January 19 (see page 215)

Late winter and early spring
> January 20 to March 31 (page 47)

Spring and early summer
> April 1 to June 17 (page 97)

High summer
> June 18 to September 9 (page 129)

Autumn
> September 10 to November 19 (page 180)

The Regional Climates of Britain

1 South-west England and Channel Islands

The south-western region is taken to include Cornwall, Devon, Somerset, Gloucestershire, Dorset and the western portion of Wiltshire. This area is largely dominated by its proximity to the sea, although the northern and eastern portion of the region often experiences rather different weather. In many respects the closeness of the Atlantic means that the weather resembles that encountered in the west of Ireland or the Hebrides. Generally, the climate is extremely mild, although that in the Scilly Isles is drier, sunnier and much milder than the closest part of the Cornish peninsula, just 40 km farther north. In the prevailing moist, south-westerly airstreams, the islands are not only surrounded by the sea, but they are fairly flat with no hills to cause the air to rise and produce rain. The Channel Islands, well to the east, are affected by their proximity to France and sometimes come under the influence of anticyclonic high-pressure conditions on the near Continent, so their overall climate tends to be more extreme.

Despite its generally mild weather, the region has experienced extremes, such as the exceptional snowfall in March 1891 that paralysed southern counties and the British rainstorm record held by Martinstown in Dorset.

Most of the peninsula of Devon and Cornwall sees very few days of frost and some areas are almost completely frost-free. Temperatures are lower, of course, over the high ground of Bodmin Moor and Dartmoor. Indeed, those areas and the Mendip Hills and Blackdown Hills do all have slightly different climatic regimes. The influence of the Severn Estuary extends well inland, and actually has an effect on the weather in the Midlands (see page 18). In winter, it allows mild air to penetrate far inland. In Cornwall and Devon, particularly in summer, sea breezes from opposite sides of the peninsula converge over the high ground that runs along the centre of the peninsula, leading to the formation of major cumulonimbus clouds and frequent showers, which may give extreme rainfall. It was this that led to the Lynmouth disaster in August 1952, when waters from a flash flood devastated the town and caused the deaths of 34 people. A somewhat similar situation arose in August 2004 in nearby

Channel Islands

Alderney

St Peter Port
Guernsey   Sark

St Helier
Jersey

Gloucester   Cheltenham

Stroud

Swindon
BRISTOL   Kingswood
Clevedon   Chippenham
Weston-super-Mare   Bath
Frome   Trowbridge

Barnstaple   Bridgwater   Salisbury

Taunton   Yeovil

SOUTH WEST

Ferndown   Christchurch
Poole
Exeter   Bournemouth
Exmouth
Newton   Weymouth
Abbot   Torquay
St Austell   Paignton
Truro   Plymouth
Penzance
Falmouth

SCILLY
ISLANDS

Boscastle and Crackington Haven, although in that instance, no lives were lost.

The escarpment of the Cotswolds, overlooking the Severn valley, often proves to be a boundary between different types of weather. This is particularly the case when there is a north-westerly wind. Then, heavy showers may affect areas on the high ground, while it is warmer and with less wind over the flatter land in the Severn valley and around Gloucester. It is often bitterly cold on the high ground above the escarpment.

2 South-east England and East Anglia

The weather in the south-eastern corner of the country may be divided into two main areas: the counties along the south coast (Hampshire, West and East Sussex and Kent); and the Home Counties around London and East Anglia, although East Anglia (Norfolk and Suffolk in particular) often experiences rather different conditions to the Home Counties.

The coastal strip from Hampshire (including the Isle of Wight) eastwards to southern Kent has long been recognised as the warmest and sunniest part of the British Isles. This largely arises from the longer duration of warm tropical air from the Continent when compared with the length of time that such air penetrates to more northern areas. The coastal strip from Norfolk to northern Kent does experience some warming effect when the winds are in the prevailing south-westerly direction, offsetting the cold North Sea. This coast may experience severe weather when there is an easterly or north-easterly airflow over the North Sea. This is particularly the case in winter: cold easterly winds bring significant snowfall to the region. It is also a feature of spring and early summer when temperatures are reduced with an onshore wind off the North Sea.

Frosts and frost hollows are a feature of the South and North Downs, the Chiltern Hills (in Berkshire, Bedfordshire and Hertfordshire) and in the high ground in East Anglia. Here, the chalk subsoil loses heat by night as do the sandy soils of Surrey and Breckland in Norfolk, leading to ground frost in places in any month of the year.

Along the south coast there is a tendency for most rain to fall in the autumn and winter, whereas for the rest of the region (the Home Counties and East Anglia) precipitation tends to occur more-or-less equally throughout the year. Because of the reliance upon groundwater, the whole region sometimes suffers from drought, when winter rains (in particular) have been insufficient to recharge the underground reserves.

3 The Midlands

The Midlands region consists of a very large number of counties: Shropshire, Herefordshire, Worcestershire, Warwickshire, West Midlands, Staffordshire, Nottinghamshire, Lincolnshire, Leicestershire, Rutland, Northamptonshire and the southern part of Derbyshire (excluding the high ground of the High Peak in the north). Of all the regions of the British Isles, this is naturally the area that has the least maritime influence. The region has been likened to a shallow bowl with surrounding hills (the Welsh Marches, the Cotswolds, the Northamptonshire Escarpment, the Derbyshire Peak and the Staffordshire Moorlands) and with a slight dome in the centre (the Birmingham Plateau). In winter, the warmest area is that closest to the Severn Valley, where warm south-westerlies may penetrate inland, whereas in summer the warmest region lies to the north-east, farthest from those moderating winds. Yet the western area is also prone to very cold nights in autumn, winter and early spring. (The lowest temperature ever recorded in England was -26.1°C at Newport in Shropshire on 10 January 1982.) Frosts are a feature of the whole region, partly because of the sandy nature of most of the soils and also because of the lack of maritime influence.

Precipitation is fairly evenly spread across the region, although the west, along the Welsh border and the high northern area of the High Peak has the highest rainfall. The east (Lincolnshire and the low ground in the east of Nottinghamshire and Northamptonshire along the valleys of the Trent and Nene) tends to be drier. Because of the rain-shadow created by the Welsh mountains, over which considerable rainfall occurs, some areas of the west of this Midlands region are drier than might otherwise

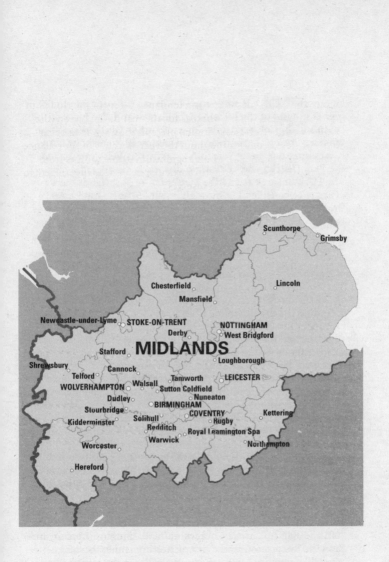

be expected. There is some increase in rainfall over the slightly higher ground of the Birmingham Plateau and also towards the south and the Cotswold hills. Towards East Anglia there is a strong tendency for most rain to occur in summer, when showers are most numerous. In the very hilly areas on the Welsh border and in the Peak District, the wettest months are December and January. The Derbyshire Peak and the Staffordshire Moorlands tend to experience considerable snowfall in winter, as do high areas of the Welsh Marches. On the lower ground to the east, snowfall is greater in the year than in the west. This is particularly the case when there are easterly or north-easterly winds that penetrate inland and bring snow from the North Sea.

4 North-west England and the Isle of Man

The north-west region consists of the land west of the Pennine chain, that is Cumbria and Lancashire in particular, especially including the mountainous Lake District in Cumbria, but also extends south to include Merseyside, Greater Manchester, Cheshire and the western side of Derbyshire. The region's weather tends to be mild and wetter in winter than regions to the east of the Pennines, and cooler in summer than regions to the south. It was, of course, the mild, relatively damp climate that was responsible for the region being the centre for the spinning of cotton, in contrast to the wool handled in the drier east. The maritime influence is seen in the fact that coastal areas are often warmer in winter and cooler in summer than areas farther inland.

There is a great difference in the amount of precipitation between the north and south of this region. The north, in Cumbria and the Lake District is notorious for high rainfall. Seathwaite, in the Lake District, currently holds the record for rainfall in 24 hours and is the wettest inhabited location in Britain. The extreme rainfall has often contributed to severe flooding, such as that in 2005 and 2009 in Carlisle, Cockermouth, Workington, Appleby and Keswick. By contrast, rainfall is much less over the Cheshire plain, which like much of Merseyside, actually lies in the rain shadow of the Welsh mountains and is thus much drier.

The prevailing wind from the south-west may give very high wind speeds over the high ground of the Pennines, while an

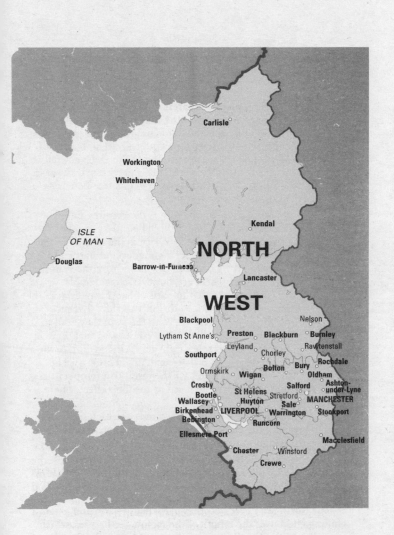

ISLE OF MAN

Douglas

NORTH

WEST

Carlisle

Workington

Whitehaven

Kendal

Barrow-in-Furness

Lancaster

Blackpool

Nelson

Lytham St Anne's

Preston

Blackburn

Burnley

Leyland

Chorley

Rawtenstall

Southport

Bolton

Bury

Rochdale

Ormskirk

Wigan

Oldham

Crosby

Salford

Ashton-under-Lyne

Bootle

St Helens

Stretford

MANCHESTER

Wallasey

Huyton

Sale

Birkenhead

LIVERPOOL

Warrington

Stockport

Bebington

Runcorn

Ellesmere Port

Macclesfield

Chester

Winsford

Crewe

21

easterly wind may produce the only named British wind, the viciously strong, and noisy, Helm wind, as air cascades over the escarpment west of Cross Fell and over the Eden valley in the north of Cumbria.

Most years see some early snow in autumn on the high fells, and in the north on the high ground it may be persistent although rarely lasts throughout the winter. The low ground along the coast and in the south sees relatively little snow and what does fall remains lying for just a few days.

5 North-east England and Yorkshire

The region of north-east England is well-defined by geographical boundaries. On the west is the Pennine range, on the east, the North Sea. The northern boundary may be taken as the river Tweed and the southern as the estuary of the Humber. The region includes very high moorland in the north and west. In general, the ground slopes down from the Pennine chain towards the coast. There are, however, considerable areas of lower land, such as that in parts of Northumberland and, in particular, in the South Riding of Yorkshire. Because the prevailing winds in Britain are from the west, they tend to deposit most of their rainfall over the Pennines, so that there is a rain shadow effect that reaches right across this region to the coast, and the whole region is drier than might otherwise be expected. The mountains also have the result that temperatures to leeward (to the east) are enhanced by the föhn effect, where descending air is warmer than that at a corresponding altitude on the windward side. With westerly winds, the high fells also tend to break up the cloud cover, so that the whole area to the east is surprisingly sunny. On the other hand, the North Sea exerts a strong cooling effect, keeping general temperatures fairly low. The region is, however, open to easterly and northerly winds and tends to suffer from gales off the sea, accompanied by heavy rain or, in winter, by snow.

The North Sea never becomes particularly warm and exerts a chilling effect over the length of the region. Sea breezes, which occur because of the difference between the warm land and the cold sea, commonly set in, especially in late spring and often

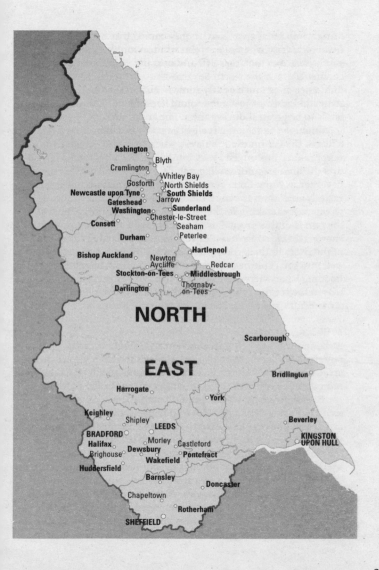

Ashington
Blyth
Cramlington
Whitley Bay
Gosforth
North Shields
Newcastle upon Tyne
South Shields
Gateshead
Jarrow
Washington
Sunderland
Chester-le-Street
Consett
Seaham
Durham
Peterlee
Bishop Auckland
Hartlepool
Newton
Aycliffe
Redcar
Stockton-on-Tees
Middlesbrough
Darlington
Thornaby-
on-Tees

NORTH

Scarborough

EAST

Bridlington

Harrogate
York

Keighley
Beverley
Shipley
LEEDS
BRADFORD
KINGSTON
Halifax
Morley
Castleford
UPON HULL
Brighouse
Dewsbury
Pontefract
Wakefield
Huddersfield
Barnsley
Doncaster
Chapeltown
Rotherham
SHEFFIELD

bring damp sea fog, or 'haar' to the coastal strip. Such conditions frequently arise when other parts of the country, especially the south-east, are enjoying settled, warm, anticyclonic weather. The cooling effect of the North Sea may be so great that there is a difference of as much as 10 degrees Celsius between the coastal strip and locations just a few kilometres inland. This difference tends to be greatest during the summer months.

Although the region has a low overall rainfall because of the effect of the Pennines to the west, when the wind is easterly it may produce prolonged heavy rainfall, as the air is forced to rise over the high ground, causing it to deposit its moisture as rain or, in winter, as snow. Long periods of heavy rain may occur from the easterly airflow on the northern side of depressions, the centres of which are tracking farther south across the country. Most of the rain is of this, frontal, type as there are few of the convective showers that produce heavy rainfall in regions farther south. Most of the rivers in the region have large catchment areas, extending well into the Pennines, and are therefore subject to episodes of flooding when there is prolonged rain. This is particularly the case in the south of the region, where the Ouse frequently floods around York and where the city of Hull often suffers.

6 Wales

Although Wales is being treated here as a single climatic region, in reality there are considerable differences in various areas. The whole country, is, of course, dominated by the long mountainous spine running from Snowdonia in the north to the Brecon Beacons, Black Mountain and the Rhondda in the south with an extension to the west in the Prescelli Mountains in Carmarthenshire and eastern Pembrokeshire. This area is not only generally high, but it is also exposed, windy and wet. To the west, the area around the whole of Cardigan Bay has a very maritime climate. It is windy, but may be particularly fine when the mountains provide shelter from easterly winds. The driest area of Wales is in the north-east, in the lowlands bordering on Cheshire. Here there is a distinct rain shadow effect in the prevailing south-westerly maritime winds. To the south of this area, the country along the border with England is also subject

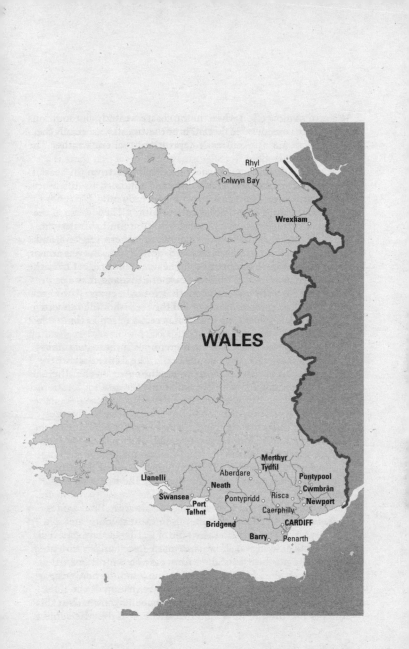

to a rain shadow effect when the winds are westerly, but may experience considerable rainfall in the anticlockwise circulation of winds around the centres of depressions that track across southern England.

The final distinct area is that of south Wales, from Pembrokeshire round to Glamorgan and southern Monmouthshire. This coastal area is generally mild, but tends to be wet and may be surprisingly cold in winter. The low-lying area around Cardiff, Newport and southern Monmouthshire is often affected in summer by fine weather over southern England, and may, at times, with southerly or south-westerly winds, even derive some shelter from Exmoor on the other side of the Bristol Channel. In summer, warm conditions in southern England may also be transported west by easterly winds. In winter, by contrast, the area may be cold as cold air drains out of the English Midlands along the Severn valley, bringing low stratus cloud or fog to the south-east of Wales.

Because of the nature of the mountainous spine to the country, both sides may sometimes be affected by the föhn effect, where descending air is warmer than the ascending air on the other side of the mountains. This is particularly the case for locations around Cardigan Bay when there are easterly winds, although the effect may be offset by the development of cooling sea breezes (or even sea mists that drift inland) that usually set in during the day. Perhaps more importantly, the föhn effect acts at sites in the north-east that are warmed when the wind is in the prevailing, south-westerly, direction. (This, of course, is one of the main reasons why the north-east is the warmest and driest area of Wales.)

Overall, the climate of Wales is a maritime one, and winter is generally mild and days in spring and summer are cool. The coastal areas all round Wales also tend to be fairly sunny, especially as convective cloud and showers are more frequently experienced inland. The north coast of the country is noted for its warm temperatures during winter, which are enhanced when there are stable southerly flows of air that may be increased by the föhn effect. The amount of rainfall varies considerably from about 700 mm per year around Rhyl on the north coast to over 4000 mm a year near Snowdon. Most of the country experiences about 1000 mm of precipitation per year. Snowfall is generally low in the coastal areas, but it may be very heavy inland when slow-moving fronts (usually moving from the south-west) encounter the high ground. However, significant falls have also taken place when easterly winds have risen over the mountain chain and deposited their moisture in the form of snow.

7 Ireland

The climate of Ireland is dominated, as might be expected, by the Atlantic or, more specifically, by the warm waters of the North Atlantic Drift off its western coast. The whole island is subject to the maritime influence and is certainly far warmer than might otherwise be expected from its latitude. This is borne out by the extent, the quantity and quality of its grasslands. The length of time when the mean temperature exceeds 6°C, which is accepted as the limit for the growth of grass is exceptionally long. It is not for nothing that Ireland has earned the nickname of the 'Emerald Isle'. The climate is exceptionally equitable and there is very little variation in temperatures throughout the year. However, its location is firmly beneath one of the major storm tracks that are followed by depressions arriving from the Atlantic. By bearing the brunt of any severe storms, not only does the island act as an 'early warning system' for the remainder of the British Isles, but also tends to temper their effects and reduce their severity for regions farther to the east.

Taken as a whole, the higher ground is located around the coasts of Ireland, with lower ground in the centre of the island. There is therefore a tendency for rainfall to be higher around the coasts than in the centre of the island. One consequence of the warm sea is that in winter, in particular, the temperature difference between the cool land and the warm sea helps to strengthen developing depressions, so that frontal systems – and the rain that they bear – tend to be stronger in winter than in summer. Rainfall on hills close to the western coasts is greater at that time of the year, when there are also more showers, which add to the overall total amount of rain. This is particularly the case when unstable polar maritime air arrives in the wake of a depression. As it passes over the main flow of the North Atlantic Drift, it becomes strongly heated from below – sometimes by as much as 9 degrees Celsius – which thus increases its instability and the strength of the showers that are generated. All of which increases the rainfall.

As far as temperatures are concerned, there is very little variation across the island, although there tends to be a greater range in the north-east when compared with the south-west. Distance from the coast also plays a part. The area with the greatest range of temperature is in southern Ulster, which experiences colder temperatures in late autumn and winter than areas along the southern and western coasts. From late spring until early autumn, maximum temperatures are higher in the north-east than in southern and western Ireland. In high summer, latitude does play a part, with temperatures being slightly higher in southern areas than along the north coast. One consequence of the equable maritime climate is that high temperatures, even in high summer, are rare, especially when compared with those that are experienced in southern England. On very rare occasions, in winter, the frigid air of the Siberian high may extend right across England and into Ireland and remains strong enough to overcome the influence of the warm air from the Atlantic. This was the case in 1962/1963 and in 2009/2010, but these winters were exceptional. Otherwise, frost is rare in coastal areas and occurs on just some 40 occasions in inland regions. As may be expected, snowfall is rare. Only in the extreme north-east does snow fall on an average of 30 days a year. In the far south-west, this figure decreases to about 5 days a year, and nowhere does snow lie for more than about a day or so.

The difference between the western coasts and the more sheltered inland areas is perhaps most obvious when wind strengths are compared. The strongest winds are observed on the northern coast of Ulster, which tends to be close to the area of the Atlantic over which depressions may undergo explosive deepening, with a consequent increase in wind speed. On rare occasions, such as the 'Night of the Big Wind' in 1839, highly destructive winds may extend right across the island, even into the far south.

8 Scotland

As with Wales, Scotland consists of several areas with diverse climates, which are a result of its mountainous as well as its maritime nature. Again, as in the case with Wales, there are five distinct areas. Because they are at a distance from the mainland, the three island areas of the Hebrides (or Western Isles), Orkney and Shetland form one climatic area. In the west, there is a distinction between the very mountainous Western Highlands region, running from Sutherland in the north, right the way down the west coast. The climate here is extremely wet because of the mountainous nature. To the east of this region, particularly in Caithness, Moray and most of Aberdeenshire, although still a highland region (the Eastern Highlands), the area is shielded from the prevailing westerly winds by the Cairngorm Mountains and warmed by the föhn effect (page 10). It is, however, fully exposed to frigid northerly winds, especially in the north, in Caithness and particularly so in winter. On the east coast farther to the south, the climate is essentially the same from the Moray Firth, down the coastal strip of Aberdeenshire, Angus and Fife, across the Firth of Forth and as far as the eastern end of the Borders. Farther inland, the Central Belt and the western area of the Borders in Dumfries and Galloway and Ayrshire form yet another climatic region.

The highest temperatures are recorded in all areas in July (sometimes July and August in the outer islands), and there is a difference of some 2–3 degrees Celsius between the average temperature recorded in the Borders in the south and that found in the extreme north of the Eastern Highlands region (in Caithness). It is striking that the highest December and January temperatures in the whole of Britain have been recorded in northern Scotland. In both cases the temperature was 18.3°C, and both occurred as a result of the föhn effect in the lee of high mountains. On 2 December 1948, Ashnaschellach in the mountainous Western Highlands region experienced this temperature and on 26 January 2003 it was Aboyne in the Eastern Highlands region in Aberdeenshire. High temperatures tend to occur in the Western Highlands with southerly or

ORKNEY
ISLANDS

SHETLAND
ISLANDS

Elgin

Inverness

Aberdeen

SCOTLAND

Perth

Dundee

Stirling

Kirkcaldy

Paisley

GLASGOW   **EDINBURGH**

Paisley   Livingston

Irvine   East Kilbride

Kilmarnock

Ayr

Dumfries

south-easterly winds, whereas the highest in the Eastern Highlands occur with westerly winds. The lowest temperatures occur in December or January in all regions with the exception of the Hebrides, Orkney and Shetland, where the lowest temperatures are recorded in February. It is believed that lower temperatures than those recorded at Braemar on 11 February 1893 and again on 10 January 1982 and at Altnahara on 30 December 1990 may have occurred at other locations, where there are no recording stations.

Scotland is known for its wet climate. There is, however, quite a striking difference between rainfall in the west (about 1250 mm per year in the outer islands and even more in the Western Highlands region) and rainfall in the eastern coastal region (that running from the Moray Firth down to the Borders). Here, yearly totals of just 650–750 mm are typical. This key feature of the Scottish climate is, of course, related to the mountainous nature of the land on the west, which receives most of the precipitation and shields the rest of the country. Precipitation also falls as snow and, once again, there is a distinct difference between the west and east of the country. Snowfall is also strongly dependent on altitude and here again the Western Highlands region received a greater amount of snow than elsewhere. In the Western Isles and particularly in western Ayrshire, days on which snow is lying are very few.

Clouds

Altocumulus clouds.

Altocumulus

Altocumulus clouds (Ac) are medium-level clouds, with bases
at 2–6 km, that, like all other varieties of cumulus, occur as
individual, rounded masses. Although they may appear in small,
isolated patches, they are normally part of extensive cloud
sheets or layers and frequently form when gentle convection
occurs within a layer of thin altostratus (page 34) breaking it up
into separate heaps or rolls of cloud. The individual elements
have apparent widths of between 1° and 5°, so their sizes lie
between those of the higher, tiny cirrocumulus (page 35) and the
lower, much larger stratocumulus (page 41). Altocumulus clouds
may also take on the appearance of flat 'pancakes', but whatever
the shape of the individual cloudlets, they always show some
darker shading, unlike cirrocumulus. Blue sky is often visible
between the separate masses of cloud – at least in the nearer
parts of the layer. These clear lanes occur where the convection
currents are descending.

Layers of altocumulus move as a whole, carried by the general wind at their height, but wind shear often causes the cloudlets to become arranged in long rolls or billows, which usually lie across the direction of the layer's motion. High altocumulus or cirrocumulus of this type give rise to beautiful clouds that are commonly known as 'mackerel skies'.

Altostratus

Altostratus (As) is a dull, medium-level white or bluish-grey cloud in a relatively featureless layer, which may cover all or part of the sky. When illuminated by the rising or setting Sun, gentle undulations on the base may be seen, but these should not be confused with the regular ripples that often occur in altocumulus (page 33).

As with stratus, altocumulus may be created by gentle uplift. This frequently occurs at a warm front, where initial cirrostratus thickens and becomes altostratus, and the latter may become rain-bearing nimbostratus (pages 39–40). Patches or larger areas of altostratus may remain behind fronts, shower clouds or larger, organised storms. Conversely, altostratus may break up into altocumulus. Convection may then eat away at the cloud, until nothing is left.

Altostratus.

Cirrus.

Cirrus

Cirrus (Ci) is a wispy, thread-like cloud that normally occurs high in the atmosphere. Usually white, it may seem grey when seen against the light if it is thick enough.

Cirrus consists of ice crystals that are falling from slightly denser heads where the crystals are forming. In most cases, wind speeds are higher at upper levels, so the heads move rapidly across the sky, leaving long trails of ice crystals behind them. Occasionally, the crystals fall into a deep layer of air moving at a steady speed. This can produce long, vertical trails of cloud.

Cirrocumulus

Cirrocumulus (Cc) is a very high white or bluish-white cloud, consisting of numerous tiny tufts or ripples, occurring in patches or larger layers that may cover a large part of the sky. The individual cloud elements are less than 1° across. They are sometimes accompanied by fallstreaks of falling ice crystals.

Unlike altocumulus (page 33), the small cloud elements do not show any shading. They are outlined by darker regions where the clouds are very thin or completely missing. These darker areas are

Cirrocumulus.

where air is descending around the edges of the convection cells, causing the cloud to disperse.

The cloud layer is often broken up into a regular pattern of ripples and billows. Clouds of this sort are often called a 'mackerel sky', although the term is sometimes applied to fine, rippled altocumulus. In fact, the differences between cirrocumulus and altocumulus are really caused only because the latter are lower and thus closer to the observer.

Cirrostratus
Cirrostratus (Cs) is a thin sheet of ice-crystal cloud and is most commonly observed ahead of the warm front of an approaching depression, when it is often the second cloud type to be noticed, after individual cirrus streaks. On many occasions, however, cirrostratus occurs as such a thin veil that it goes unnoticed, at least initially, until one becomes aware that the sky has lost its deep blue colour and has taken on a slightly milky appearance. The Sun remains clearly visible (and blindingly bright) through the cloud, but as the cirrostratus thickens, a slight drop in temperature may become apparent.

Once you realise that cirrostratus is present, it pays to check the sky frequently, because as it thickens it may display striking halo phenomena. This stage often passes fairly quickly as the cloud continues to thicken and lower towards the surface, eventually turning into thin altostratus (page 34).

Cirrostratus often has a fibrous appearance, especially if it arises from the gradual increase and thickening of individual cirrus streaks. Because the cloud is so thin, and contrast is low, the fibrous nature is easier to see when the Sun is hidden by lower, denser clouds or behind some other object.

A solar halo in thin, almost invisible, cirrostratus cloud.

Cumulus.

Cumulus

Cumulus clouds (Cu) are easy to recognise. They are the fluffy clouds that float across the sky on a fine day, and are often known as 'fair-weather clouds'. The individual heaps of cloud are generally well separated from one another – at least in their early stages. They have rounded tops and flat, darker bases. It is normally possible to see that these bases are all at one level. Together with stratus (page 40) and stratocumulus (page 41) they form closer to the ground than other cloud types.

The colour of cumulus clouds, like that of most other clouds, depends on where they are relative to the Sun and the observer. When illuminated by full sunlight, they are white – often blindingly white – but when seen against the Sun, unless they are very thin, they are various shades of grey.

Cumulonimbus

Cumulonimbus (Cb) is the largest and most energetic of the cumulus family. It appears as a vast mass of heavy, dense-looking cloud that normally reaches high into the sky. Its upper portion is usually brilliantly white in the sunshine, whereas its lower portions are very dark grey. Unlike the flat base of a cumulus,

Cumulonimbus.

the bottom of a cumulonimbus is often ragged and it may even reach down to just above the ground. Shafts of precipitation are frequently clearly visible.

Cumulonimbus clouds consist of enormous numbers of individual convection cells, all growing rapidly up into the sky. Although cumulonimbus develop from tall cumulus, a critical difference is that at least part of their upper portions has changed from a hard 'cauliflower' appearance to a softer, more fibrous look. This is a sign that freezing has begun in the upper levels of the cloud.

Nimbostratus

Nimbostratus (Nb) is a heavy, dark grey cloud with a very ragged base. It is the main rain-bearing cloud in many frontal systems. Shafts of precipitation (rain, sleet or snow) are visible below the cloud, which is often accompanied by tattered shreds of cloud that hang just below the base.

Just as cirrostratus often thickens and grades imperceptibly into altostratus, so the latter may thicken into nimbostratus. Once rain actually begins, or shafts of precipitation are seen to reach the ground, it is safe to call the cloud nimbostratus.

Nimbostratus.

Stratus

Stratus (St) is grey, water-droplet cloud that usually has a fairly
ragged base and top. It is always low and frequently shrouds
the tops of buildings. Indeed, it is identical to fog, which may
be regarded as stratus at ground level. Although the cloud

Stratus.

may be thin enough for the outline of the Sun to be seen clearly through it, in general it does not give rise to any optical phenomena. It forms under stable conditions and is one of the cloud types associated with 'anticyclonic gloom'.

Stratus may form either by gentle uplift (like the other stratiform clouds) or when nearly saturated air is carried by a gentle wind across a cold surface, which may be either land or sea. Normally, low wind speeds favour its occurrence, because mixing is confined to a shallow layer near the ground. When there is a large temperature difference between the air and the surface, however, stratus may still occur, even with very strong winds. Stratus also commonly forms when a moist air-stream brings a thaw to a snow-covered surface.

There is very little precipitation from stratus, but it may produce a little drizzle or, when conditions are cold enough, even a few snow or ice grains. Ragged patches of stratus, called 'scud' by sailors, often form beneath rain clouds such as nimbostratus or cumulonimbus, especially where the humid air beneath the rain cloud is forced to rise slightly, such as when passing over low hills.

Stratocumulus

Stratocumulus (Sc) is a low, grey or whitish sheet of cloud, but unlike stratus it has a definite structure. There are distinct, separate masses of cloud that may be in the form of individual clumps, broader 'pancakes', or rolls. Sometimes these may be defined by thinner (and thus whitish) regions of cloud, but frequently blue sky is clearly visible between the masses of cloud. The latter are always more than 5° across (roughly the width of three fingers, held at arm's length). If they appear smaller, the cloud is classed as altocumulus.

Stratocumulus indicates stable conditions, and only slow changes to the current weather. It generally arises in one of two ways: either from the spreading out of cumulus clouds that reach an inversion, or through the break-up of a layer of stratus cloud. In the first case, the tops of the cumulus flatten and spread out sideways when they reach the inversion, producing clouds that are fairly even in thickness, with flat tops and bases. Initially, perhaps early in the day, there may be large areas of clear air, but the individual elements gradually merge to cover a larger area, or even completely blanket the sky.

In the second case, shallow convection (whose onset is often difficult to predict) begins within a sheet of stratus, causing the layer to break up. The regions of thinner cloud or clear air indicate where the air is descending, and the thicker, darker centres where it is rising.

Stratocumulus.

Cloud heights

The height of clouds is usually given in feet (often with approximate metric equivalents). This may seem odd, when all other details of clouds, and meteorology in general, uses metric (SI) units. It is, however, a hang-over from the way in which aircraft heights are specified. When aviation became general between the two World Wars, most commercial flying took place in the United Kingdom and North America, so heights of aircraft and airfields were given in feet. It was obviously essential for cloud heights to be the same. The practice has continued: the heights of airfields, aircraft and clouds are still given in feet. The World Meteorological Organization recognises three ranges of cloud heights: low, middle and high. Clouds are specified by the height of their bases, not by that of their tops. The three divisions are:

Low clouds (bases 6500 feet or lower, approx. 2 km and below): cumulus (page 38), stratocumulus (page 41), stratus (page 40).

Middle clouds (bases between 6500 and 20,000 feet, approx. 2 to 6 km): altocumulus (page 33), altostratus (page 34), nimbostratus (page 39).

High clouds (bases over 20,000 feet, above 6 km): cirrus (page 35), cirrocumulus (page 35), cirrostratus (page 36).

One cloud type, cumulonimbus (page 38), commonly stretches through all three height ranges. Nimbostratus, although nominally a middle-level cloud, is frequently very deep and although it has a low base, may extend to much higher altitudes.

January

Introduction

January sees some of the coldest temperatures of the year. Braemar, the village in the Scottish highlands, about 93 km west of Aberdeen, has twice recorded the lowest temperature in the British Isles (-27.2°C), first on 11 February 1895 and since then on 10 January 1982. Only Altnaharra in Sutherland has ever recorded a similar temperature (on 30 December 1995). Generally, Scotland sees extensive snow cover in January, particularly important at the ski centres at Cairngorm Mountain, Glencoe Mountain, Glenshee, Nevis Range and The Lecht, although global warming threatens to decrease the coverage of snow. Deep snow has become less frequent and the centres have had to invest in snow-making cannons.

The weather in January is often dominated by cold easterly winds. These arise because of the cold anticyclone over the near continent that builds up during most winters. This is often an extension of the great wintertime Siberian High that dominates the weather over Asia and creates a cold, dry airflow over the eastern side of Asia. Circulation round this anticyclone tends to bring a cold airflow across the North Sea. In crossing the sea, the air gains moisture and this often results in snowfall along the eastern coast of Britain. At times, a blocking situation may arise with a major high-pressure area over Scandinavia. This brings extremely cold Arctic air down over the British Isles and, depending how long the block persists, may give rise to a persistent spell of low temperatures.

Snowfall is, however, very frequent farther south, and the exceptional snowfall in 1947, which was undoubtedly the winter that saw the greatest fall of snow over Britain, did not begin until quite late in the month of January. Although there had been some snow earlier, in December and early January, it had melted by the middle of the month, and there were unseasonably high temperatures across the country. The temperature then dropped and there were frosts at night from January 20 (the nominal beginning of the late winter, early spring season). Snow-bearing clouds began to move into the south-west of England on January 22. There was heavy snowfall with blizzard conditions in the West Country. Even the Scilly Isles saw a slight covering of snow, amounting to a few centimetres in depth – an almost unprecedented event. The following days saw heavy snowfall extend right across all of England and Wales before spreading into Scotland. There were seemingly relentless snowstorms over the next few weeks, which left England and Wales, up as far as the Scottish Borders, buried beneath a deep blanket of snow, and movement by rail and road almost completely paralysed. Somewhat ironically, although snow fell somewhere in the United Kingdom on every day from January 22 until the middle of March, Scotland escaped the worst storms. After some three months of northerly and easterly winds, by March 10, warm southerly and south-westerly winds started to affect the West Country. Not only did the warm airstream bring dense fogs, it also produced heavy rainfall, which in turn led to floods as the rain ran off the frozen ground. The situation was worsened by the gradual thawing of the immense snowpack, leading to even more extreme flooding. By March 13, even the rivers in East Anglia were about to burst their banks.

Only the winter of 1962–63 saw a longer period during which snow persisted, and much lower temperatures, but the amount of snow that fell in 1947 was the most extreme ever recorded for the United Kingdom.

Late winter and early spring season – January 20 to March 31
This season tends to exhibit long spells of settled conditions. These may be of very cold weather, characterised by Arctic air, introduced by a northerly airflow, and thus forming the main period of winter. Although in some years the weather may take on the character of an extended spring, such conditions are less frequent than those with low temperatures. There may be long spells of wet, westerly conditions, but these tend to be less common than spells with cold northerly or easterly winds. However, the wet westerlies suddenly become less frequent after about March 9, and indeed westerly weather then becomes very uncommon.

Weather Extremes

Country	Temp.	Location	Date
Maximum temperature			
England	17.6°C	Eynsford (Kent)	27 Jan. 2003
Northern Ireland	16.4°C	Knockarevan (Co. Fermanagh)	26 Jan. 2003
Scotland	18.3°C	Aboyne (Aberdeenshire) Inchmarlo (Kincardineshire)	26 Jan. 2003
Wales	18.3°C	Aber (Gwynedd)	10 Jan. 1971 27 Jan. 1958
Minimum temperature			
England	-26.1°C	Newport (Shropshire)	10 Jan. 1982
Northern Ireand	-17.5°C	Magherally (Co. Down)	1 Jan. 1979
Scotland	-27.2°C	Braemar (Aberdeenshire)	10 Jan. 1982
Wales	-23.3°C	Rhayader (Powys)	21 Jan. 1940

Country	Pressure	Location	Date
Maximum pressure			
Scotland	1053.6 hPa	Aberdeen Observatory	31 Jan. 1902
Minimum pressure			
Scotland	925.6 hPa	Ochtertyre (Perthshire)	26 Jan. 1884

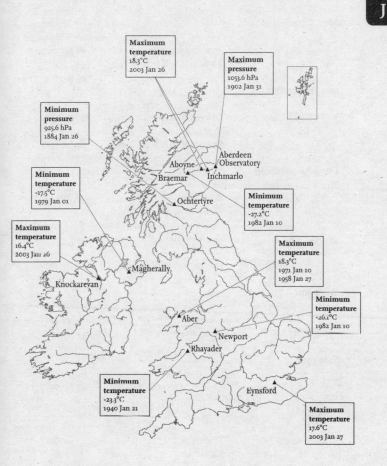

Maximum
temperature
18.3°C
2003 Jan 26

Maximum
pressure
1053.6 hPa
1902 Jan 31

Minimum
pressure
925.6 hPa
1884 Jan 26

Minimum
temperature
-17.5°C
1979 Jan 01

Aberdeen
Observatory
Aboyne
Braemar   Inchmarlo

Ochtertyre

Minimum
temperature
-27.2°C
1982 Jan 10

Maximum
temperature
16.4°C
2003 Jan 26

Maximum
temperature
18.3°C
1971 Jan 10
1958 Jan 27

Knockarevan

Magherally

Minimum
temperature
-26.1°C
1982 Jan 10

Aber

Newport

Rhayader

Minimum
temperature
-23.3°C
1940 Jan 21

Eynsford

Maximum
temperature
17.6°C
2003 Jan 27

The Weather in January 2021

Observation	Location	Date
Max. temperature		
14.2°C	Pershore College (Hereford & Worcester)	28 January
Min. temperature		
-13.0°C	Dawyck Botanic Garden (Peeblesshire)	9 January
	Braemar (Aberdeenshire)	31 January
Overnight minimum		
10.3°C	St Mary's (Scilly Isles) Westonzoyland (Somerset)	19–20 January
24-hour rainfall		
132.8 mm	Honister Pass (Cumbria)	20 January
Wind gust		
60 knots (69 mph)	South Uist (Western Isles) Marham (Norfolk)	15 January 21 January
Sunshine		
8.1 hours	Shoeburyness (Essex)	25 January
Snow depth		
32 cm	Trassey Slievenaman (Co. Down)	25 January

J

The year began with a cold northerly airflow, which gradually turned to the north-east. The airflow brought Arctic air and cold conditions that affected Scotland and northern England, with snow in Scotland and on the high ground of England and Wales. The north-west of Scotland experienced heavy snow cover and cold temperatures. The conditions caused travel disruptions in Scotland, northern England and Northern Ireland.

The weather in the south of the country turned warmer and wetter around January 7, when there was freezing fog in the Midlands, but colder conditions persisted in the north.

Very wet and windy weather arrived with Storm Christoph between January 19 and 21, causing major floods in Greater Manchester and Cheshire (in particular), heavy rainfall in Cumbria, and considerable snow falls in parts of Scotland as the storm cleared the country. It caused a landslip in the Rhondda valley in Wales. There was flooding in south-west England and even in the east.

Mild conditions then covered the south-west, giving unseasonably warm temperatures, with particularly high temperatures over the night of January 19 to 20. Following the transit of storm Christoph, it was generally colder, with snow in the West Country and in the Midlands. The last week of the month saw the weather turn much milder and wet in the south of England, but it remained cold in Scotland.

Sunrise and Sunset 2022

Location	Date	Rise	Azimuth	Set	Azimuth
Belfast					
	Jan 01 (Sat)	08:46	131	16:09	229
	Jan 11 (Tue)	08:41	128	16:23	232
	Jan 21 (Fri)	08:30	125	16:40	236
	Jan 31 (Mon)	08:15	120	17:00	241
Cardiff					
	Jan 01 (Sat)	08:18	128	16:14	233
	Jan 11 (Tue)	08:15	125	16:27	235
	Jan 21 (Fri)	08:06	122	16:43	238
	Jan 31 (Mon)	07:53	117	17:00	243
Edinburgh					
	Jan 01 (Sat)	08:43	133	15:49	228
	Jan 11 (Tue)	08:38	130	16:04	230
	Jan 21 (Fri)	08:26	126	16:23	234
	Jan 31 (Mon)	08:09	121	16:44	239
London					
	Jan 01 (Sat)	08:07	128	16:02	232
	Jan 11 (Tue)	08:03	125	16:15	235
	Jan 21 (Fri)	07:54	122	16:31	238
	Jan 31 (Mon)	07:41	118	16:48	243

Note that all times are in Universal Time (UT), otherwise known as Greenwich Mean Time (GMT). These times do not take Summer Time (BST) into account.

Moonrise and Moonset 2022

Location	Date	Rise	Azimuth	Set	Azimuth
Belfast					
	Jan 01 (Sat)	07:41	137	14:25	222
	Jan 11 (Tue)	12:18	67	02:24	290
	Jan 21 (Fri)	20:40	73	10:27	291
	Jan 31 (Mon)	08:33	136	15:34	226
Cardiff					
	Jan 01 (Sat)	07:08	132	14:36	226
	Jan 11 (Tue)	12:15	68	02:05	288
	Jan 21 (Fri)	20:35	74	10:08	290
	Jan 31 (Mon)	08:01	132	15:42	229
Edinburgh					
	Jan 01 (Sat)	07:40	139	14:03	220
	Jan 11 (Tue)	12:03	66	02:16	291
	Jan 21 (Fri)	20:25	72	10:20	292
	Jan 31 (Mon)	08:32	138	15:12	224
London					
	Jan 01 (Sat)	06:56	133	14:23	226
	Jan 11 (Tue)	12:03	68	01:53	288
	Jan 21 (Fri)	20:22	74	09:56	290
	Jan 31 (Mon)	07:50	132	15:29	229

Note that all times are in Universal Time (UT), otherwise known as Greenwich Mean Time (GMT). These times do not take Summer Time (BST) into account.

Twilight Diagrams 2022

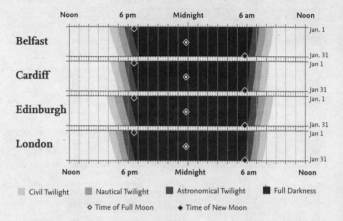

The exact times of the Moon's major phases are shown on the diagrams opposite.

In 1854, **Robert Fitzroy** was appointed to the new post of Meteorological Statist to the Board of Trade. The office was the forerunner of the modern Met Office. Fitzroy is noted for developing ways of warning sailors of impending storms and for his study of the *Royal Charter* storm of 1859 (pages 208–209).

Robert Fitzroy was a pioneer in meteorology, who provided predictions of forthcoming weather. He introduced the term 'forecast', but suffered considerable criticism because his methods were considered not to be sufficiently scientific and based on a knowledge of how the weather would develop.

The Moon's Phases and Ages 2022

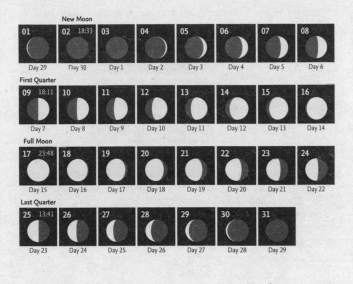

William Lindley Wragge (generally known as Clement Wragge) climbed Ben Nevis (1345 metres) every day between June and the middle of October 1881 to make meteorological observations from the peak.

Clement Wragge was appointed Government Meteorologist for Queensland in 1887. He became heavily involved in the study of the tropical cyclones that affect Queensland, and was the first person to give specific names to these systems.

On This Day

1 January 1928 – Thomas Baird and Hugh Barrie, caught in a blizzard on Cairngorm, either fell or were killed by an avalanche, becoming the first official hillwalking deaths.

7 January 1839 – A violent storm damaged the suspension bridge across the Menai Strait that had been built by Thomas Telford.

8 January 1982 – An extreme snowstorm raged for 40 hours, creating 6-metre drifts in Wales, closing roads and isolating many places. The snow persisted for days and some remote locations were not reached until January 19. Although Wales was hardest hit, many other places, particularly in the West Country, were also affected.

8 January 2005 – Carlisle was flooded when the rivers Eden, Petterill, Caldew and Little Caldew overtopped their banks. More than 3000 houses were flooded.

13 January 1805 – Francis Beaufort first described his famous wind scale in the log for HMS *Woolwich*.

J

15 January 1867 – Forty people drowned when the ice on the lake in Regent's Park, London, broke under the weight of skaters.

16 January 1930 – In thick fog, the ship *Romanie* was wrecked in Polridmouth Cove, Cornwall. The author, Daphne du Maurier, re-created the shipwreck in her famous novel *Rebecca*.

18 January 2007 – A gust of 99 mph (159 kph) was recorded at the Needles on the Isle of Wight.

21 January 1876 – Snow jammed a railway signal at Abbots Ripton, near Huntingdon, Cambridgeshire, leading to a triple train crash in which 13 people were killed and 59 injured.

The Loss of the *Princess Victoria*

January 1953 saw a number of weather-related disasters. The one that seems to have captured the general imagination is the occurrence of the disastrous east coast floods on 31 January 1953. It is true that disaster caused the death of 307 people in England, 19 in Scotland and many more – 1,826 – deaths in the Netherlands. That particular event led to major changes in forecasting, and also to the construction of flood barriers across the Thames and the River Hull in Britain and major works – especially the Delta Plan – in the Netherlands. It was not the only tragedy. A major disaster that happened on the same day and that has been largely 'forgotten' was the sinking of the ro-ro (roll-on/roll-off) ferry *Princess Victoria* in the North Channel as it was crossing from Ayrshire to Northern Ireland, and the resulting loss of 135 lives. This was the worst single incident of the events caused by an extremely deep depression and the associated high winds.

The ferry left the protected waters of Loch Ryan in Ayrshire at 07:45 and proceeded to sea. Its stern was exposed to the worst of the storm waves. A separate 'guillotine' door had been fitted as a result of doubts over the integrity of the standard stern doors, but this additional door had not been lowered (because it was difficult to do). The stern doors became damaged, and the ship started to take on water and develop a list to starboard. Attempts to turn back were frustrated and a decision was made to continue to Larne in Northern Ireland, shaping a course to protect the stern doors as far as possible. Because of the deteriorating situation, a request for assistance was made at 09:46 and an SOS transmission at 10:32. There was confusion over the location of the vessel, because the words 'not under command' were taken to mean that way had been lost. So the coastguard (incorrectly) estimated the ship's position and acted accordingly. In fact, the vessel was still heading towards

J

Northern Ireland. The order to abandon ship was given at 14:00. Although ships arrived shortly afterwards, they were unable to rescue survivors because of the extreme danger of the lifeboats being dashed against the sides of the larger vessels. Only when actual, purpose-built lifeboats, in particular the RNLI lifeboat, the *Sir Samuel Kelly* from Donaghadee, arrived from Northern Ireland was it possible to rescue any individuals. Forty-four lives were saved, but including none of the ship's officers. The ship's radio operator was posthumously awarded the George Cross for remaining at his post, thus allowing passengers and crew to escape.

A number of other vessels were in distress at this time, and several trawlers were also lost. The subsequent enquiry into the disaster found that some of the assistance that might have helped the *Princess Victoria*, particularly aircraft assistance, was already helping with other emergencies caused by the same storm.

Rain shadow

An area to the leeward of high ground, whether hills or mountains, often experiences lesser rainfall than neighbouring areas or than expected. This rain-shadow effect occurs because as air rises over the higher ground (usually to the west) there is increased rainfall, leaving less moisture to fall on any areas to leeward of the hills.

February

Introduction

Although in many rural areas (in southern England in particular), February has earned the nickname 'February Filldyke', in reality February may be extremely dry. February, is, in fact, the one calendar month that is most likely to experience no rainfall whatsoever. As such, it sometimes shows spring-like conditions, which warrants its inclusion in the 'late winter, early spring' season (page 47). It has been found that a warm February in Scotland is a sign that the mean annual temperature in Scotland will probably be warm. However, a warm February may not be welcome for agriculture. Warmth indicates that precipitation will be in the form of rain, rather than snow. Rain in February is generally detrimental to seed and grass, but snow protects the ground and keeps it warm.

Yet spells of very cold weather may occur, especially if, as is often the case in January, a blocking situation occurs, especially a blocking anticyclone over Scandinavia, bringing frigid air down from the Arctic or similarly cold air from the semi-permanent cold anticyclone that builds up over Siberia in winter. The latter was the case with the exceptionally cold spell in 2018, nicknamed 'The Beast from the East' by the media, which began late in the month on February 22, and continued until at least the end of the month. This particular cold wave originated in an exceptional anticyclone, named Anticyclone Hartmut, which transported frigid Siberian air west over Europe. In crossing the North Sea, the air gained a large amount of humidity, which produced extremely heavy snowfall that spread west over almost the whole of the United Kingdom and Ireland.

Somewhat similar conditions occurred in the middle of March (March 17 and 18), although this second cold spell did not last particularly long, and was not as severe. (The media sometimes called it the 'Mini Beast from the East'.)

The weather in February often shows a division between the north and south of the country, with cold temperatures, and relatively dry conditions, prevailing in the north of England and over Scotland, but much milder, wetter weather in the south, sometimes with exceptional rainfalls – hence the 'February Filldyke' nickname. The 'Beast from the East' just mentioned not only brought cold conditions to Britain, but the frigid air it brought from the east then had a great effect on the humid air introduced from the west by Storm Emma. (This storm was actually named by the meteorological services of France, Spain and Portugal, rather than the name being taken from the list prepared by the Met Office and Met Eireann.)

Storm Emma originated in the Azores and was a typical deep depression, transporting a lot of warm moist air and accompanied by high winds. When the storm encountered the frigid air introduced over the British Isles by Anticyclone Hartmut, the warm, moist air was forced up over the cold air at the surface and Emma's moisture was deposited as snowfall. The snow depth reached as much as 57 cm in places, although a depth of 50 cm was widespread across the country. South-west England and south Wales were worst affected. Storm Emma also brought a renewed incursion of Arctic air and the resulting low temperatures over much of the United Kingdom.

The effects of this collision between Hartmut and Emma were not confined to the British Isles. Snow fell along the French Riviera and in Italy. Even Barcelona in Spain saw snowfall – an unheard-of event for the region. The collision of the two systems also produced some exceptional winds, most notably a gust of 228 kph (142 mph) at Mont Aigoual in southern France on 1 March 2018.

Weather Extremes

Country	Temp.	Location	Date
Maximum temperature			
England	19.7°C	Greenwich Observatory	13 Feb. 1998
Northern Ireland	17.8°C	Bryansford (Co. Down)	13 Feb. 1998
Scotland	17.9°C	Aberdeen (Aberdeenshire)	22 Feb. 1897
Wales	18.7°C	Colwyn Bay (Conwy)	23 Feb. 2012
Minimum temperature			
England	-20.6°C	Woburn (Bedfordshire)	25 Feb. 1947
Northern Ireland	-15.0°C	Armagh (Co. Armagh)	7 Feb. 1895
Scotland	-27.2°C	Braemar (Aberdeenshire)	11 Feb. 1895
Wales	-20.0°C	Welshpool (Powys)	2 Feb. 1954

Country	Pressure	Location	Date
Maximum pressure			
Scotland	1052.9 hPa	Aberdeen (Aberdeenshire)	1 Feb. 1902
Minimum pressure			
Eire	942.3 hPa	Midleton (Co. Cork)	4 Feb. 1951

F

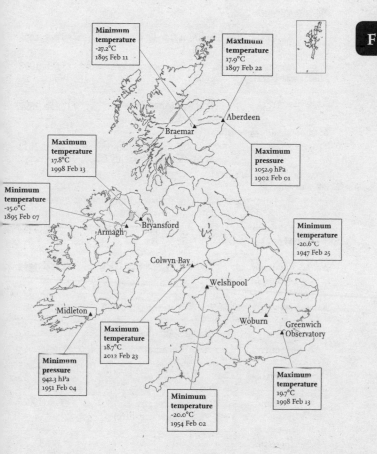

Minimum
temperature
-27.2°C
1895 Feb 11

Maximum
temperature
17.9°C
1897 Feb 22

Maximum
temperature
17.8°C
1998 Feb 13

Minimum
temperature
-15.0°C
1895 Feb 07

Maximum
pressure
1052.9 hPa
1902 Feb 01

Minimum
temperature
-20.6°C
1947 Feb 25

Minimum
pressure
942.3 hPa
1951 Feb 04

Maximum
temperature
18.7°C
2012 Feb 23

Minimum
temperature
-20.0°C
1954 Feb 02

Maximum
temperature
19.7°C
1998 Feb 13

Aberdeen

Braemar

Bryansford

Armagh

Colwyn Bay

Welshpool

Midleton

Woburn

Greenwich
Observatory

The Weather in February 2021

Observation	Location	Date
Max. temperature		
18.4°C	Stanton Downham (Suffolk)	24 February
Min. temperature		
-23.0°C	Braemar (Aberdeenshire)	11 February
Max. overnight temperature		
12.2°C	St Mary's Airport (Scilly Isles)	23 February
24-hour rainfall		
125.8 mm	Honister Pass (Cumbria)	24 February
Wind gust		
72 knots (83 mph)	South Uist (Western Isles)	14 February
Snow depth		
38 cm	Aboyne (Aberdeenshire)	10 February

The beginning of February 2021 saw relatively mild conditions in the south of the country, but cold in the north. There was heavy rain and some flooding in Devon and Cornwall on February 1–2, and snow brought disruption to Northern Ireland on February 2. Heavy snow then affected northern England on February 3 and in Scotland there were road and rail closures because of snow on February 5–6. Temperatures over England dropped after February 6 and there was then considerable snowfall in the eastern counties. The worst disruption arose when an easterly wind brought heavy snowfall to blanket the eastern coastal counties of England, particularly East Anglia and Kent, on February 7–8. Up to 10 cm of lying snow was reported from some locations. There was major disruption to travel by road and rail and even in London some services were cancelled.

Farther north there was major snowfall in northern England (particularly in the northern Pennines) and in Scotland, with extremely heavy falls in the Scottish mountains, with some communities being cut off by deep snow. An avalanche risk was even declared for the Pentland hills, south of Edinburgh. Lying snow to a depth of 26 cm was reported in County Durham on February 11. In the south-west, when the rest of the country was trying to cope with deep snow, there were strong south-easterly winds, which disrupted rail services at Dawlish in Devon.

In Northern Ireland, a combination of strong winds, snow and heavy rain proved disruptive to both road and rail travel on February 13–14. High winds were recorded in Scotland, with a gust of 72 knots (83 mph) recorded at South Uist in the Hebrides on February 14. In northern England, the winds caused problems by driving drifting snow. Heavy rain then spread from the south-west to reach all parts of England by nightfall on February 14. Showery rain then affected most of England, with even isolated thunderstorms on February 16. A band of heavy rain swept across the country on February 18 and strong winds and gales arose in exposed locations.

February 19 and 20 saw very heavy rain affecting the west of the country, with road and rail closures in Northern Ireland, Scotland, Wales, the western counties of the Midlands and Devon and Cornwall in the south-west. There was widespread flooding in Wales, where the rivers Cynin, Towy and Usk all flooded. Even farther east there were problems, with flooding in Hertfordshire. Devon and Cornwall continued to suffer disruption on February 21. The whole month saw high rainfall and some locations in the north-east of England and Cumbria had double their normal rainfall.

Heavy rain continued to affect the whole of England, especially the north, and on February 23, Honister Pass in Cumbria (a notoriously wet location) saw 125.8 mm of rain, although farther south and east it was much brighter, with an overnight temperature of 11.2°C at St Mary's Airport in the Scilly Isles. There were also higher temperatures in the south-east on 24 February, with the month's maximum being reported at Stanton Downham in Suffolk, but still heavy rain in the west and north.

The month ended with brighter conditions over much of England, particularly in the south and south-east. Indeed, many southern locations recorded surprisingly long hours of sunshine in the last three days of the month.

F

Alexander Buchan (1829–1907) was a Scottish meteorologist who, after an initial career as a teacher, became Secretary of the Scottish Meteorological Society in 1860, and remained in that position for 47 years, until his death.

Buchan is most famous for his proposed 'Buchan spells'. He claimed that colder or warmer periods occurred every year and were predictable. Although the 'spells' were truly only applicable to weather in Scotland, they were seized upon by the media, and became a vogue applied to the whole country. Nowadays, meteorologists consider that the 'spells' are purely random, and do not exist.

Buchan is credited with the invention of proper weather forecasting. He used surface synoptic charts to follow North American storms as they crossed the Atlantic and affected the British Isles.

Alexander Buchan, photographed some time before 1897. (From the Fridtjof Nansen Archive, National Library of Norway.)

Sunrise and Sunset 2022

Location	Date	Rise	Azimuth	Set	Azimuth
Belfast					
	Feb 01 (Tue)	08:13	119	17:02	241
	Feb 11 (Fri)	07:54	113	17:23	247
	Feb 21 (Mon)	07:32	107	17:44	253
	Feb 28 (Mon)	07:16	103	17:58	258
Cardiff					
	Feb 01 (Tue)	07:51	117	17:02	243
	Feb 11 (Fri)	07:34	112	17:20	248
	Feb 21 (Mon)	07:15	106	17:39	254
	Feb 28 (Mon)	07:00	102	17:51	259
Edinburgh					
	Feb 01 (Tue)	08:07	120	16:46	240
	Feb 11 (Fri)	07:47	114	17:08	246
	Feb 21 (Mon)	07:24	108	17:30	253
	Feb 28 (Mon)	07:07	103	17:45	257
London					
	Feb 01 (Tue)	07:40	117	16:50	243
	Feb 11 (Fri)	07:23	112	17:08	248
	Feb 21 (Mon)	07:03	106	17:27	254
	Feb 28 (Mon)	06:49	102	17:39	258

Note that all times are in Universal Time (UT), otherwise known as Greenwich Mean Time (GMT). These times do not take Summer Time (BST) into account.

Moonrise and Moonset 2022

Location	Date	Rise	Azimuth	Set	Azimuth
Belfast					
	Feb 01 (Tue)	09:04	129	17:07	234
	Feb 11 (Fri)	11:44	42	05:03	317
	Feb 21 (Mon)	23:58	113	09:19	253
	Feb 28 (Mon)	07:04	133	14:35	230
Cardiff					
	Feb 01 (Tue)	08:37	126	17:10	237
	Feb 11 (Fri)	11:54	46	04:30	313
	Feb 21 (Mon)	23:38	111	09:14	255
	Feb 28 (Mon)	06:35	129	14:41	233
Edinburgh					
	Feb 01 (Tue)	09:01	130	16:48	232
	Feb 11 (Fri)	11:22	40	05:02	319
	Feb 21 (Mon)	23:51	114	09:05	253
	Feb 28 (Mon)	07:02	135	14:14	228
London					
	Feb 01 (Tue)	08:26	126	16:57	237
	Feb 11 (Fri)	11:42	46	04:18	313
	Feb 21 (Mon)	23:25	111	09:01	255
	Feb 28 (Mon)	06:23	129	14:28	233

Note that all times are in Universal Time (UT), otherwise known as Greenwich Mean Time (GMT). These times do not take Summer Time (BST) into account.

Twilight Diagrams 2022

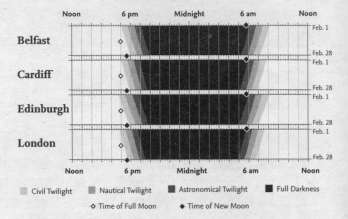

The exact times of the Moon's major phases are shown on the diagrams opposite.

Jet streams

Jet streams are narrow ribbons of fast moving air, typically hundreds of kilometres wide and a few kilometres in depth. The most important one for British weather is the Polar Front Jet Stream, a westerly wind that flows right round the Earth. It is driven by the great temperature difference between the cold polar air and warmer air closer to the equator. Fluctuations in latitude are primarily caused by the flow across the Rockies in North America. It affects the strength of depressions and also their paths, causing them to sometimes pass across the British Isles and sometimes to the north or south of them.

The Moon's Phases and Ages 2022

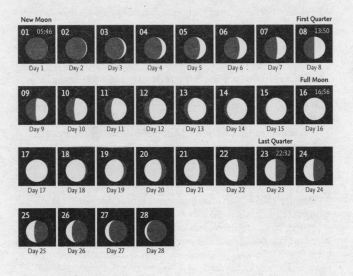

New Moon 01 05:46

First Quarter 08 13:50

Full Moon 16 16:56

Last Quarter 23 22:32

Day 1	Day 2	Day 3	Day 4	Day 5	Day 6	Day 7	Day 8
Day 9	Day 10	Day 11	Day 12	Day 13	Day 14	Day 15	Day 16
Day 17	Day 18	Day 19	Day 20	Day 21	Day 22	Day 23	Day 24
Day 25	Day 26	Day 27	Day 28				

F

Rear-Admiral Sir Francis Beaufort (1774–1857), was a British naval officer who in 1806 devised a means of estimating wind strength at sea. His scheme was not adopted by the Royal Navy until 1838. Beaufort's scale (including an adaptation for use on land) is still in use today (pages 246–249).

In 1829, Beaufort was appointed head of the Admiralty's Hydrographic Office, a post that he held for 25 years. Under his leadership, the Office became the world's leading hydrographic organisation. Beaufort made major contributions to many scientific fields, including geography, geodesy, oceanography, and astronomy, as well as meteorology.

On This Day

2 February 1901 – Queen Victoria's funeral takes place on a cold, gloomy day in London.

5 February 1941 – In a south-easterly gale, the SS *Politician* was wrecked on the Scottish island of Eriskay. The ship was carrying 25,000 cases of whisky. The event inspired Compton Mackenzie, in 1946, to produce the tale of the SS *Cabinet Minister*, wrecked on the island of Great Toddy, and the basis of the famous film *Whisky Galore!*

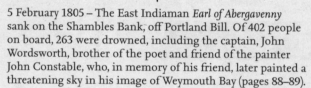

5 February 1805 – The East Indiaman *Earl of Abergavenny* sank on the Shambles Bank, off Portland Bill. Of 402 people on board, 263 were drowned, including the captain, John Wordsworth, brother of the poet and friend of the painter John Constable, who, in memory of his friend, later painted a threatening sky in his image of Weymouth Bay (pages 88–89).

6 February 1963 – The blizzard began that deposited the most snow of the phenomenal winter of 1962–63.

7 February 1969 – A gust of 118 knots (136 mph) was recorded at Kirkwall in Orkney.

F

11 February 1895 – Braemar in Aberdeenshire reported the coldest temperature ever known in Britain, -27.2°C. This stood as the record until 10 January 1982, when Braemar again recorded the same temperature.

16 February 1962 – Sheffield suffered from an exceptional windstorm. In a north-westerly gale, waves created by the Pennines to the west of the city were trapped and amplified by a temperature inversion above the mountains, resulting in winds of 80 mph (130 kph) at the surface. At least 100,000 homes were damaged to a greater or lesser extent and some (about 100) were completely destroyed.

21 February 1861 – The spire of Chichester Cathedral, built by Sir Christopher Wren, collapsed, having been battered by winds from the north-east the day before, and then by gale-force south-westerly winds.

28 February 1662 – Known as 'Windy Tuesday'. It was the strongest storm in England between 'St Maury's Wind' of 1362 and the Great Storm of 1703.

The North Sea Floods of 1953

Although generally associated with the date of 31 January 1953, most of the flooding, destruction and deaths actually occurred the next day, 1 February 1953. The very deep depression that had caused the winds and high seas that led to the loss of the *Princess Victoria* ferry (page 58) swung round the north of Scotland, with the centre passing between Orkney and Shetland. It then moved down the length of the North Sea, and headed in the general direction of the Netherlands. The low-pressure centre was accompanied on its western side by strong northerly and north-westerly winds, which were gale force or severe gale force.

The effects of the winds, the low pressure (which always tends to raise the sea surface) and a high spring tide combined to create a storm surge and raise water levels as much as 5.6 metres above normal in various locations along the North Sea coast. The rise was particularly great in the southern North Sea. The high sea level overtopped the flood defences almost everywhere along the coast, with most of the flooding occurring after midnight. Damage occurred to some 1,600 km of the British coastline. There were devastating floods in East Anglia and around the Thames Estuary, with some 30,000 people force to flee from their homes, and there was damage to about 24,000 properties.

The problem was not confined to the southern North Sea, and there were as many as 19 fatalities on the east coast of Scotland. Casualties in Lincolnshire, Norfolk, Suffolk and Essex amounted to 307. There were 28 deaths in north-east Belgium, but a devastating 1836 deaths in the Netherlands, in the province of Zeeland, parts of the province of South Holland, and north-western part of the province of North Brabant. In both the United

Kingdom and the Netherlands, the situation was exacerbated by the fact that it occurred on the night of Saturday to Sunday. All local government offices were closed at the weekend and no warnings of any sort were issued of the dangerous conditions.

One of the effects of this devastating flood was that better arrangements were put in place for distributing warnings to the general populace. In addition, meteorologists also created a system for determining the likelihood of storm surges and ensuring that the relevant warnings were issued to the appropriate authorities. The event was the main stimulus for the construction of the Thames Barrier at Woolwich on the Thames Estuary. (Although London was lucky to escape major flooding on this particular occasion.) A tidal barrier was also constructed at the mouth of the River Hull, where it enters the estuary of the Humber. Both barriers have proved their worth on subsequent occasions. In the Netherlands the event led to the development of the complex Delta Works (commonly known as the Delta Plan) to seal the openings of the Rhein, Meuse and Scheldt estuaries, preventing any surge from propagating inland and also to shorten the length of the coastline to be strengthened and raised.

A somewhat similar surge on 16–17 February 1962 affected northern Germany, and gave rise to extreme flooding in Hamburg, even though the city is at least 100 km away from the coast. There, 315 people lost their lives and as many as 60,000 houses were damaged. The same windstorm created extensive damage in the city of Sheffield (page 75).

March

Introduction

March has traditionally been viewed as a transitional month between winter and spring. It has come to be associated with the saying 'In like a lion, out like a lamb'. Regrettably for this saying, although the weather may well be changeable, there is no distinct pattern that applies every year. The month may well begin with a settled period, with little in the way of winds and dramatic events, and then deteriorate. Interestingly, there is often a sharp change in the weather after the first week of March. The frequency of westerly weather – that is, the progression of depressions across the country from the west – drops suddenly, and very few are recorded. The number of such depressions and sequences of weather is less than at any other time of the year, except for late April and early May. In general, the increasing strength of heating from the Sun begins to take effect from the beginning of the month. Although there may still be spells of very cold weather, with a lot of cloud, the increased warmth is readily noticed.

Another tradition, strongly held by some, particularly by mariners, is that March experiences 'equinoctial gales'. (The equinox is on March 20 in 2022.) This idea is, however, not supported by the evidence. Gales are actually most common around the winter solstice (December 21 in 2022), and least frequent around the summer solstice (June 21). There may be a perception that gales are most frequent around the autumnal equinox (September 23 in 2022), but this is probably a result of the deterioration of the weather from the fine, long days of summer to the shorter, less settled days of autumn. When the Sun is well north of the equator, its warmth generally causes the Azores High to strengthen and extend its influence as ridges of high pressure. These may reach so far as to cover the British Isles and most of western Europe. Sometimes the high-pressure region extends well to the west, giving rise to what is known as the Bermuda High. Depressions, and their accompanying cloud and winds, tend to follow the Polar Front jet stream, which is diverted towards the north. The low-pressure systems, with their clouds, rain and winds, pass north of the British Isles, which thus remains under the influence of the settled, anticyclonic weather within the extended Azores High. With the change of season from summer to autumn, the Sun starts to move south. It moves fastest towards the equator and south of it in September, usually causing the high-pressure area over the Azores to decline. When this happens, depressions are no longer displaced northwards and they, and their accompanying winds, thus cross the British Isles more directly. It would seem that mariners' observations of such windier weather in September have been taken in a more general sense to suggest that gales occur at both equinoxes (in March and September). There is no evidence whatsoever for increased windiness at either equinox, so this is yet another instance of incorrect weather lore.

M

Gulf Stream

A warm-water current on the western side of the North Atlantic Ocean. It extends along the eastern seaboard of the United States from the Gulf of Mexico to Cape Hatteras. It then turns eastwards and becomes the North Atlantic Current. The warm water affecting the British Isles is a branch of this current, known as the North Atlantic Drift (often incorrectly called the Gulf Stream). This branch leaves the main current in mid-Atlantic and, passing west of Ireland, heads up towards Norway and the Arctic Ocean.

Weather Extremes

Country	Temp.	Location	Date
Maximum temperature			
England	25.6°C	Mepal (Cambridgeshire)	29 Mar. 1968
Northern Ireland	21.7°C	Armagh (Co. Armagh)	28 Mar. 1965 29 Mar. 1965
Scotland	23.6°C	Aboyne (Aberdeenshire)	27 Mar. 2012
Wales	23.0°C	Prestatyn (Denbighshire) Ceinws (Powys)	29 Mar. 1965
Minimum temperature			
England	-21.1°C	Houghall (Co. Durham)	4 Mar. 1947
Northern Ireland	-14.8°C	Katesbridge (Co. Down)	2 Mar. 2001
Scotland	-22.8°C	Logie Coldstone (Aberdeenshire)	14 Mar. 1958
Wales	-21.7°C	Corwen (Denbighshire)	3 Mar. 1965

Country	Pressure	Location	Date
Maximum pressure			
England	1047.9 hPa	St Mary's Airport (Isles of Scilly)	9 Mar. 1953
Minimum pressure			
Scotland	946.2 hPa	Wick (Caithness)	9 Mar. 1896

M

Minimum pressure
946.2 hPa
1896 Mar 09

Minimum temperature
-22.8°C
1958 Mar 14

Maximum temperature
23.6°C
2012 Mar 27

Wick

Logie Coldstone • Aboyne

Minimum temperature
-14.8°C
2001 Mar 02

Minimum temperature
-21.1°C
1947 Mar 04

Maximum temperature
21.7°C
1965 Mar 28
1965 Mar 29

Houghall

Armagh • Katesbridge

Maximum temperature
23.0°C
1965 Mar 29

Prestatyn
Corwen

Ceinws • • Mepal

Maximum temperature
23.0°C
1965 Mar 29

Maximum temperature
25.6°C
1968 Mar 29

St Mary's Airport •

Maximum pressure
1047.9 hPa
1953 Mar 09

Minimum temperature
-21.7°C
1965 Mar 03

The Weather in March 2021

Observation	Location	Date
Max. temperature		
24.5°C	Kew Gardens (Greater London)	30 March
Min. temperature		
-8.5°C	Braemar (Aberdeenshire)	6 March
Max. overnight temperature		
12.7°C	Wych Cross (East Sussex)	30 March
24-hour rainfall		
177.7 mm	Seathwaite (Cumbria)	28 March
Wind gust		
86 knots (99 mph)	Needles Old Battery (Isle of Wight)	13 March
Snow depth		
3 cm	Tulloch Bridge (Inverness-shire)	13 March
Sunshine		
12.1 hours	Shoeburyness (Essex)	30 March

The start of the month was relatively quiet, with generally anticyclonic conditions affecting the country. There was no dramatic weather anywhere in the British Isles, although it was generally cold, with Braemar yet again recording a very low temperature. This changed dramatically on March 9 with extremely high winds, particularly in the north-western Highlands of Scotland, where there were school closures and power cuts to many thousands of homes. Northern Ireland experienced high winds and transport difficulties on March 10 and 11. Wales was also affected with heavy rain and winds, with speed restrictions or closure of some bridges, together with power cuts in some areas. Both north-west and north-east England suffered from high winds and flooding. High winds also affected the Midlands and southern England, with disruption to both road and rail services in London and the south-east.

In the middle of the month, heavy rain and winds affected Wales on March 10 and showers and winds then spread across most of England, with some hail and even thunder on March 12. A period of patchy, light rain then followed, although it was warm on the eastern side of the country, with especially high temperatures in the east of Scotland. Heavy rain then spread from the north-west to most of the country. Practically all of the British Isles experienced cloudy conditions on March 20 and 21, although there were some sunny intervals.

On March 26 and 27 it was generally colder, with some snow on hills in the north. Although the south of England was fine, there was very heavy rain in the north-west of England on March 28, with Seathwaite in Cumbria recording as much as 177.2 mm of rainfall. Some rain persisted the following day in the north, but the southern counties became warm and dry, and this situation tended to persist to the end of the month.

At the end of the month, heavy rain affected north-west Scotland on March 28 and 29, causing flooding in places. In contrast, the south of England was very sunny and warm, with only slight overnight frost and long periods of sunshine.

M

Sunrise and Sunset 2022

Location	Date	Rise	Azimuth	Set	Azimuth
Belfast					
	Mar 01 (Tue)	07:13	102	18:00	258
	Mar 11 (Fri)	06:49	95	18:20	265
	Mar 21 (Mon)	06:24	88	18:39	272
	Mar 31 (Thu)	05:59	82	18:58	279
Cardiff					
	Mar 01 (Tue)	06:58	101	17:53	259
	Mar 11 (Fri)	06:36	95	18:10	265
	Mar 21 (Mon)	06:13	89	18:27	272
	Mar 31 (Thu)	05:51	82	18:44	278
Edinburgh					
	Mar 01 (Tue)	07:04	102	17:47	258
	Mar 11 (Fri)	06:39	95	18:08	265
	Mar 21 (Mon)	06:13	88	18:28	272
	Mar 31 (Thu)	05:46	81	18:49	279
London					
	Mar 01 (Tue)	06:46	101	17:41	259
	Mar 11 (Fri)	06:24	95	17:59	265
	Mar 21 (Mon)	06:02	89	18:16	272
	Mar 31 (Thu)	05:39	82	18:33	278

Note that all times are in Universal Time (UT), otherwise known as Greenwich Mean Time (GMT). These times do not take Summer Time (BST) into account.

Moonrise and Moonset 2022

Location	Date	Rise	Azimuth	Set	Azimuth
Belfast					
	Mar 01 (Tue)	07:28	124	16:06	239
	Mar 11 (Fri)	10:17	40	03:56	320
	Mar 21 (Mon)	23:12	120	07:38	246
	Mar 31 (Thu)	06:16	98	18:01	267
Cardiff					
	Mar 01 (Tue)	07:03	122	16:07	242
	Mar 11 (Fri)	10:29	44	03:21	315
	Mar 21 (Mon)	22:48	118	07:36	248
	Mar 31 (Thu)	06:02	98	17:50	267
Edinburgh					
	Mar 01 (Tue)	07:24	126	15:49	238
	Mar 11 (Fri)	09:54	38	03:56	322
	Mar 21 (Mon)	23:06	121	07:23	245
	Mar 31 (Thu)	06:06	98	17:48	267
London					
	Mar 01 (Tue)	06:52	122	15:54	241
	Mar 11 (Fri)	10:16	44	03:09	315
	Mar 21 (Mon)	22:36	118	07:24	248
	Mar 31 (Thu)	05:50	98	17:38	267

M

Note that all times are in Universal Time (UT), otherwise known as Greenwich Mean Time (GMT). These times do not take Summer Time (BST) into account.

Twilight Diagrams 2022

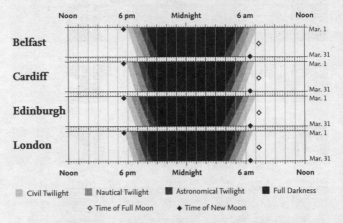

Noon	6 pm	Midnight	6 am	Noon

Belfast — Mar. 1 / Mar. 31
Cardiff — Mar. 1 / Mar. 31
Edinburgh — Mar. 1 / Mar. 31
London — Mar. 1 / Mar. 31

Civil Twilight   Nautical Twilight   Astronomical Twilight   Full Darkness
◇ Time of Full Moon   ◆ Time of New Moon

The exact times of the Moon's major phases are shown on the diagrams opposite.

The Moon's Phases and Ages 2022

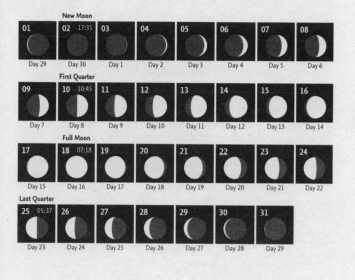

The painter, John Constable, who was well versed in meteorology and cloud studies, sometimes included weather phenomena in his paintings as a tribute to notable individuals. He included a 'meteorologically impossible' rainbow in *Salisbury Cathedral from the Meadows* as a tribute to his friend Archdeacon John Fisher, and the dark clouds (rather than 'cotton-wool' fair-weather cumulus clouds) in *Weymouth Bay: Bowleaze Cove and Jordon Hill* in memory of John Wordsworth, who was drowned in a shipwreck off Weymouth on 5 February 1805 (see page 74).

Previous page: Weymouth Bay: Bowleaze Cove and Jordon Hill *by John Constable, painted about 1816. National Gallery, London.*

On This Day

2 March 1948 – A Sabena Dakota aircraft on a flight from Brussels to London misses the runway in thick fog and crashes. Although three passengers are rescued, all three crew and 16 of the passengers die in the resulting fire. It is the first (and worst) accident at London's Heathrow airport.

8 March 1935 – Blue-black rain, caused by Saharan dust, carried high in the atmosphere, falls on Shetland during a heavy thunderstorm.

10 March 1891 – A violent, heavy snowstorm in the West Country introduces a new word to the English language. It is the very first event to be described by the word 'blizzard'.

12 March 1744 – A strong gale in the English Channel prevents a French invasion fleet, inspired by the attempt by James III to claim the throne of England and Scotland, from leaving Dunkirk.

14 March 1964 – Properly organised and equipped mountain rescue teams are introduced after three young participants (19, 21 and 24 years old) die of exhaustion and hypothermia when taking part in the 'Four Inns Walk', held annually by Derby Scouts in the High Peak area of Derbyshire.

15 March 1789 – The very first, purpose-built lifeboat, the *Original*, is constructed by Henry Greathead, as a result of a public competition launched after the loss of the *Adventure* and all its crew off South Shields on this day in March 1789.

19 March 1969 – On 19 March 1969, during strong winds, the Emley Moor television mast collapsed under the weight of the ice on the tower and guy wires. At the time of its construction in 1966, it was one of the tallest structures in the world at 385 metres.

M

20 March 2010 – The Icelandic stratovolcano, known as Eyjafjallajökull, began erupting. It had last erupted in 1823. Although the eruption began in March, the main event, which occurred under the glacier and produced a vast cloud of volcanic ash in the atmosphere, disrupting air travel, did not occur until 14–18 April 2021.

27 March 1980 – A ferocious windstorm in the North Sea caused the *Alexander Kielland* accommodation platform to collapse off the Scottish coast. Some 123 workers were drowned and an enormous international rescue effort was launched, involving nearly 50 vessels, 27 helicopters and two planes. Eventually, 89 men were rescued, proving the value of the use of rescue helicopters.

The Collapse of the Emley Moor Mast

In 1956, a lattice television transmission mast (135 metres high) was built at Emley Moor in West Yorkshire where the ground surface is at an altitude of some 260 metres above sea level. The village of Emley and the site of the mast is roughly halfway between the main conurbations of Huddersfield and Wakefield. Transmissions began on 3 November 1956. In 1966, the lattice mast was dismantled and replaced by a taller mast, with an overall height of 385 metres. At the time, it was one of the tallest structures in the world. Transmissions began almost immediately from the new mast. This second mast consisted of a cylindrical steel tube (275 metres tall and 2.75 metres in diameter), above which was a lattice structure 107 metres high and a final capping cylinder, carrying the antennae and transmission equipment. The whole structure was braced by a number of guy wires. In winter, the mast and stay wires often became coated in ice. This commonly built up to such an extent that large blocks of ice in the form of icicles became detached and fell from the guy wires. This happened so frequently that apart from the standard aircraft warning lights, red lights and warning notices were installed at a low level to indicate the hazard to drivers and pedestrians on the ground. Some nearby roads were even closed on occasions when the risk was considered to be most severe.

On 19 March 1969 there was a strong wind and very cold conditions. Ice began to accumulate on the tower and stay wires. Blocks of ice started to fall from the guy wires, so the low-level warning lights were switched on. Then, all of a sudden, the whole tubular mast collapsed. Although one broken guy wire sliced through the nearby chapel and the transmitter buildings at the base of the mast, no one was injured. Although it was initially thought that the collapse was caused by the build-up of ice on the wire stays, the subsequent enquiry suggested that an additional factor in the collapse was an oscillation that built up in the tower under a low, sustained wind speed. Appropriate modifications to prevent oscillations, including hanging steel chains weighing 150 tonnes inside the cylinders, were made to two similar masts elsewhere in the country (at Belmont in the London area and Winter Hill

in Lancashire) and these have remained standing.

The collapse caused several million homes to lose television programmes. Various temporary masts were rushed to the site over the following days and weeks and enabled services to be restored in a piecemeal fashion. The mast was later replaced on a slightly different site. The latest mast is a tapering, reinforced concrete tower, 275 metres high, surmounted by a 55-metre steel lattice. This gives the tower a total height of 330.4 metres. It is the highest freestanding structure in the United Kingdom, and a Grade II listed building, considered to be of architectural merit. The top of the tower is 594 metres above sea level. Officially, this latest tower is called the Arqiva Tower, after the name of its owners, but it is known to the general public (because of its history) as the Emley Moor Tower. One tower, the Skelton transmission tower in Cumbria is nearly 35 metres higher. This is a guyed mast and the tallest structure in the United Kingdom, with a height of 365 metres.

The third, current, Emley Moor television tower (formally the Arqiva Tower). A tapering concrete construction, 330.4 metres high, and a prominent landmark. It is a Grade II Listed building, considered to be of architectural merit.

M

Introduction

April is traditionally associated with a changeable month and the general view is that it is accompanied by 'April showers'. Although it is certainly changeable, with the transition from winter to the warmer conditions of summer, in recent years the showers have been less frequent. Certainly, there are normally periods when northerly winds (in particular) bring showers and these may be squally, and can sometimes even turn into thunderstorms and be accompanied by hail. Generally nowadays, the showers are infrequent, and it is one of the quietest times of the year. April is actually one of the driest months for most of the country, followed in sequence, by March, June, May and February.

The temperature of the sea surrounding the British Isles remains low, while the land begins to warm up. Sea breezes are common and they are often accompanied by sea fog. This is particularly frequent on the east coast of Britain. Sea fogs often form over the cold North Sea and then are carried inland by the sea breezes that build up during the day. In Scotland, in particular, the 'haar', as it is known, often moves in during the afternoon. It is often particularly noticeable when it invades the Firth of Forth and hides the bottom of the Forth Bridges from view. A similar phenomenon occurs anywhere along the East Coast of England, particularly in Northumbria, but is found as far south as East Anglia. The same sort of sea fog does occur on the western coasts of Britain, but is less common, because the sea surface temperature of the Atlantic water tends to be higher than that of the North Sea, so sea fog is less likely to form.

A

Spring and early summer season – April 1 to June 17
The weather during this season is very changeable. Indeed, it is
the most changeable of the year. There are occasional outbreaks
of northerly winds, which tend to produce heavy squally showers,
often turning into thunderstorms with lightning and even hail.
Initially, high pressure and dry air over Continental Europe often
extends west over the British Isles, but this high pressure and its
accompanying dry, continental air tend to collapse in early summer
and be replaced by moist maritime air as depressions move across
the country from the west, on tracks towards the Baltic or slightly
farther north, towards Scandinavia.

Weather Extremes

Country	Temp.	Location	Date
Maximum temperature			
England	29.4°C	Camden Square (London)	16 Apr. 1949
Northern Ireland	24.5°C	Boom Hall (Co. Londonderry)	26 Apr. 1984
Scotland	27.3°C	Inverailort (Highland)	17 Apr. 2003
Wales	26.2°C	Gogerddan (Ceredigion)	16 Apr. 2003
Minimum temperature			
England	-15°C	Newton Rigg (Cumbria)	2 Apr. 1917
Northern Ireland	-8.5°C	Killylane (Co. Antrim)	10 Apr. 1998
Scotland	-13.3°C	Braemar (Aberdeenshire)	11 Apr. 1917
Wales	-11.2°C	Corwen (Denbighshire)	11 Apr. 1978

Country	Pressure	Location	Date
Maximum pressure			
Scotland	1044 hPa	Eskdalemuir (Dumfriesshire)	11 Apr. 1938
Minimum pressure			
Northern Ireland	952.9 hPa	Malin Head (Co. Donegal)	1 Apr. 1948

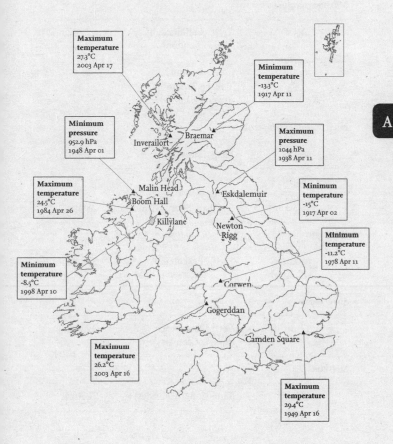

Maximum temperature
27.3°C
2003 Apr 17

Minimum temperature
-13.3°C
1917 Apr 11

Minimum pressure
952.9 hPa
1948 Apr 01

Maximum pressure
1044 hPa
1938 Apr 11

Maximum temperature
24.5°C
1984 Apr 26

Minimum temperature
-15°C
1917 Apr 02

Minimum temperature
-11.2°C
1978 Apr 11

Minimum temperature
-8.5°C
1998 Apr 10

Maximum temperature
26.2°C
2003 Apr 16

Maximum temperature
29.4°C
1949 Apr 16

Braemar

Inverailort

Malin Head

Boom Hall

Killylane

Eskdalemuir

Newton Rigg

Corwen

Gogerddan

Camden Square

A

The Weather in April 2021

	Observation	Location	Date
Max. temperature			
	21.4°C	Treknow (Cornwall)	1 April
Min. temperature			
	-9.4°C	Tulloch Bridge (Inverness-shire)	12 April
Max. overnight temperature			
	-8.4°C	Shap (Cumbria)	11 April
24-hour rainfall			
	28.8 mm	St Athan (South Glamorgan)	28 April
Wind gust			
	65 knots (75 mph)	Fair Isle (Shetland)	5 April
Snow depth			
	12 cm	Loch Glascarnoch (Ross & Cromarty)	7 April
Sunshine			
	14.2 hr	Tiree (Inner Hebrides)	30 April

Overall, the month of April 2021 was marked by anticyclonic weather, resulting in conditions for most of the country being cold and dry, but also sunny. Although mean maximum temperatures were close to the average for recent years, minimum temperatures were noticeably lower everywhere except in western coastal districts. Indeed, April 2021 proved to be colder with more air frosts than the preceding month of March. Most of the country was very dry, with only parts of northern Scotland and an area in Wiltshire having more than 50 per cent of average rainfall. Some central and eastern areas of England recorded just a few millimetres (1–4) of rain in the whole month and indeed much of this was a rain equivalent, because it fell as snow. There was a greater number of hours of sunshine than average, especially across southern Scotland and northern England.

There was heavy snowfall, accompanying a strong northerly wind across parts of Scotland on April 4 and 5, particularly in the north-east and Shetland. Difficult driving conditions were also reported in Wales and Northern Ireland. It turned very cold across the whole country from April 5, and the cold conditions continued for a couple of weeks, with widespread frosts. A deep depression tracked north of Scotland on April 7 to 9, giving northern Scotland the only significant rainfall anywhere during the month. A band of snow swept down across England from the north-west on April 12. One anticyclone moved across the country in the middle of the month, becoming centred over Scandinavia and a weak depression followed it, affecting most of the country. Yet another anticyclone moved in from the north-west, giving improved weather during the day, although remaining cold at night. It was slightly milder in Scotland at mid-month, but turned distinctly colder towards the end of April. Wales experienced showery conditions in the middle of the month, although it was generally milder in Northern Ireland.

In England the month ended with showery weather, including some thunderstorms and hail. There was even snow in the north, and in Northern Ireland. There were heavy showers in the south-west of England, with only the counties farther east along the south coast avoiding the showers.

A

Sunrise and Sunset

Location	Date	Rise	Azimuth	Set	Azimuth
Belfast					
	Apr 01 (Fri)	05:56	81	19:00	279
	Apr 11 (Mon)	05:31	74	19:19	286
	Apr 21 (Thu)	05:08	68	19:39	292
	Apr 30 (Sat)	04:48	63	19:56	298
Cardiff					
	Apr 01 (Fri)	05:48	82	18:46	279
	Apr 11 (Mon)	05:26	75	19:03	285
	Apr 21 (Thu)	05:05	70	19:19	291
	Apr 30 (Sat)	04:47	65	19:34	296
Edinburgh					
	Apr 01 (Fri)	05:44	81	18:51	280
	Apr 11 (Mon)	05:18	74	19:11	287
	Apr 21 (Thu)	04:53	67	19:31	293
	Apr 30 (Sat)	04:32	62	19:50	299
London					
	Apr 01 (Fri)	05:37	82	18:34	279
	Apr 11 (Mon)	05:14	75	18:51	285
	Apr 21 (Thu)	04:53	70	19:08	291
	Apr 30 (Sat)	04:35	65	19:23	296

Note that all times are in Universal Time (UT), otherwise known as Greenwich Mean Time (GMT).

Moonrise and Moonset

Location	Date	Rise	Azimuth	Set	Azimuth
Belfast					
	Apr 01 (Fri)	06:26	88	19:21	278
	Apr 11 (Mon)	12:19	52	04:30	310
	Apr 21 (Thu)	01:21	141	07:34	219
	Apr 30 (Sat)	04:54	72	19:41	294
Cardiff					
	Apr 01 (Fri)	06:16	88	19:07	277
	Apr 11 (Mon)	12:23	55	04:02	307
	Apr 21 (Thu)	00:45	136	07:47	223
	Apr 30 (Sat)	04:50	73	19:20	292
Edinburgh					
	Apr 01 (Fri)	06:15	88	19:11	278
	Apr 11 (Mon)	11:59	51	04:27	312
	Apr 21 (Thu)	01:22	143	07:10	216
	Apr 30 (Sat)	04:40	71	19:34	295
London					
	Apr 01 (Fri)	06:04	88	18:54	277
	Apr 11 (Mon)	12:10	55	03:50	307
	Apr 21 (Thu)	00:33	136	07:34	223
	Apr 30 (Sat)	04:38	73	19:08	292

A

Note that all times are in Universal Time (UT), otherwise known as Greenwich Mean Time (GMT).

Twilight Diagrams 2022

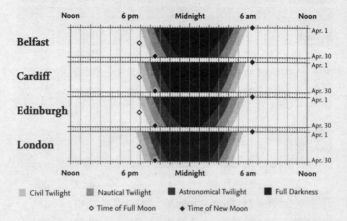

The exact times of the Moon's major phases are shown on the diagrams opposite.

Depression

A low-pressure area. (Often called a 'storm' in North-American usage.) Winds circulate around a low-pressure centre in an anticlockwise direction in the northern hemisphere. (Clockwise in the southern hemisphere.) Depressions generally move across the globe from west to east, although under certain conditions they may linger over an area or even (rarely) move towards the west.

The Moon's Phases and Ages 2022

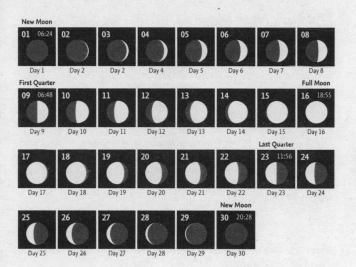

New Moon

01 06:24	02	03	04	05	06	07	08
Day 1	Day 2	Day 2	Day 4	Day 5	Day 6	Day 7	Day 8

First Quarter ... **Full Moon**

09 06:48	10	11	12	13	14	15	16 18:55
Day 9	Day 10	Day 11	Day 12	Day 13	Day 14	Day 15	Day 16

Last Quarter

17	18	19	20	21	22	23 11:56	24
Day 17	Day 18	Day 19	Day 20	Day 21	Day 22	Day 23	Day 24

New Moon

25	26	27	28	29	30 20:28
Day 25	Day 26	Day 27	Day 28	Day 29	Day 30

A

Air mass
A large volume of air that has uniform properties (particularly temperature and humidity) throughout. Air masses arise when air stagnates over a particular area for a long time. These areas are known as 'source regions' and are generally the semi-permanent high-pressure zones, which are the sub-tropical and polar anticyclones. The primary classification is based on temperature, giving Arctic (A), polar (P) and tropical (T) air.

On This Day

1 April 1875 – The first newspaper weather map was published in *The Times*. The map showed the situation on the previous day, March 31. It was prepared by Francis Galton, a fierce critic of Robert Fitzroy's empirical methods of preparing weather forecasts.

1 April 2006 – Heavy rain in the last week of March, causing a high flow in the River Severn, a strong south-westerly wind and a spring tide all combine to produce a strong Severn Bore. Steve King sets a record for surfing the bore for some 11 kilometres, but the record is not officially recognised, because the distance cannot be verified.

5 April 1815 – The initial major eruption of Mount Tambora on the island of Sumbawa.

5 April 1911 – The largest tree in England is blown down by a snowstorm in Oxford. The tree was the Huntingdon wych elm in the courtyard at Magdalen College in Oxford. The tree had a height of 43 metres, a circumference of over 8 metres and was over 400 years old.

10 April 1815 – The cataclysmic climax of the eruption of Tambora, with such widespread effects on the atmosphere that the whole world was cooled, leading to the failure of harvests the following year, and disruption to weather patterns, such that 1816 became known as 'The Year without a Summer'.

14–18 April 2010 – The Icelandic volcano Eyjafjallajökull erupted again, ejecting a large volume of fine ash into the atmosphere and ultimately leading to the great disruption of all European air traffic. The volcano was considered to have become dormant by early August.

15 April 1912 – The liner RMS *Titanic* sank in the North Atlantic, with the loss of 1517 lives. The ship struck an iceberg and was travelling fast, despite having had six warnings of sea ice.

20 April 1802 – Dorothy Wordsworth noted in her diary how during a walk on 15 April 1802, she and her brother noticed daffodils tossing in the wind. Her journal later inspires her brother William to write his most famous poem *Daffodils*.

23 April 1981 – An extreme fall of snow for late April blankets England from the northern Pennines as far south as Salisbury Plain in 8 centimetres or more of snow. The snowfall is the most severe cold spell of weather ever experienced in April in the twentieth century.

The Year without a Summer

On 5 April 1815, there was a major eruption of the volcano Tambora on the island of Sumbawa in Indonesia (or as it was then known, the Dutch East Indies). Minor activity at the volcano had started long before, in 1812. The event on April 5 was merely a precursor to the next stage, when there was a truly cataclysmic eruption of April 10, which destroyed the top of the volcano itself. It left a large caldera, some 7 kilometres across and some 600–700 metres deep. The height of the volcano was reduced by about 1500 metres. The eruption is considered to be the greatest volcanic eruption of the nineteenth century. The noise of the explosion on April 5 was definitely heard at Ternate in the Molluca Islands, some 1400 kilometres away. On April 10, what was initially thought to be distant gunfire was heard on Sumatra, at a distance of 2000 kilometres, and the most distant location at which the climactic explosion was heard was 2600 kilometres away.

The eruption sent a vast cloud of ash and (amongst other compounds) sulphur dioxide into the atmosphere. The eruption column reached a height of at least 43 kilometres, well into the stratosphere. The ash caused an immediate dimming of sunlight, especially in the immediate vicinity of the volcano, with ashfall at least 1300 kilometres away. The largest ash particles fell out fairly quickly, but the finer particles were carried right round the globe by upper atmosphere winds, and caused various optical effects. (They are believed to have caused some of the vibrant skies painted by William Turner.)

The sulphur dioxide combined with water vapour to create tiny droplets of sulphuric acid, which were also spread around the globe. The effects of the fine dust and sulphuric acid droplets, suspended in the stratosphere, were to prevent solar radiation from reaching the surface and caused the average global temperature to drop by 0.4–0.7°C. This may seem a small amount, but it had a devastating effect on agriculture and overall temperatures. It was sufficient to cause crop failures and major food shortages in the northern hemisphere. Temperatures were so low that 1816 came to be called 'The Year without a Summer', 'the Poverty Year', 'the Summer that Never Was', 'the Year There Was No Summer', and 'Eighteen Hundred and Froze to Death'.

In the eastern states of North America, the spring and early summer saw a 'dry fog', that was similar to the 'fog' that had been observed in Europe after the eruption of the Icelandic Laki volcanic chain in 1783. The 'dry fog', which was not dispersed by wind or rain, has subsequently been described as a 'sulphate veil', consisting of aerosols suspended in the stratosphere. There was widespread famine and food riots across Europe, which was recovering from the Napoleonic wars.

The disaster included effects in Britain. It is estimated that over 65,000 people died from the effects of the agricultural shortages in Britain alone. This is similar to the number of deaths in Indonesia as a direct result of the eruption, although one recent estimate suggests that these were about 95,000. Worldwide, the death toll is considered to be far more than the six million estimated to have occurred after the 1783 eruption of Laki. It is possible that the effects were increased by another volcanic eruption in 1814, but the cooling and agricultural effects persisted into 1817, with a continuation of famine conditions.

May

Introduction

The month of May has traditionally been associated with the beginning of summer, with innumerable celebrations taking place on May 1. Many of these events are very ancient festivals to mark the first day of summer.

In England, festivities on May 1 frequently included maypole dancing (often thought to be a form of fertility rite), Morris dancing, and sometimes the appearance of a Green Man. In Scotland, May Day celebrations, which have taken place for centuries, were formerly closely associated with the ancient rite of Beltane. The latter was one of the four traditional Gaelic festivals, celebrated approximately halfway between the spring equinox and the summer solstice. Over the years it had become established as May 1. In Wales, celebrations were, like those in Scotland, originally a continuation of the ancient Beltane festivities.

Although it is true that May, more than any other month, experiences more periods of high-pressure (anticyclonic) conditions, with long spells of fine, settled weather, records show that May has also experienced its fair share of extreme weather, with major snowfalls as well as extreme heat. Bank holidays for England had been established in 1871 under William Gladstone as Prime Minister. The original intention was for a bank holiday to occur at Whitsun. But Whitsun is a religious date, seven weeks after Easter and, although it often fell in May, could fall on any day between 11 May and 14 June. With such a range of dates, the weather could be extremely variable. Twenty years after the introduction of bank holidays, on the Whitsun Bank Holiday on 18 May 1891, there was a major snowfall that blanketed East Anglia. On that date, the lowest May temperature was recorded for England: -8.6°C in the notorious Rickmansworth frost hollow. Snow is not entirely uncommon this late in the year. Snow in Yorkshire was particularly deep on 17 May 1935. In mid-May 1955, the worst snowstorm for 60 years affected a large area covering Birmingham, the Cotswolds and the Chiltern Hills. On 19 May 1996, heavy snow on Dartmoor caused the annual Ten Tors adventure training weekend, run by the army, to be abandoned.

Other May weather had been very changeable. For example, that Ten Tors weekend had temperatures of 26°C in 1997, and was abandoned again in 2007, when a young girl was swept away by a brook, swollen from a depth of 1 metre to 5 metres by heavy rain. The wettest May day ever was 7 May 1881, when no less than 172.2 mm of rain fell in Cumbria. As detailed on page 124, extreme thunderstorms occurred on Derby Day, 31 May 1911. In May 1923, Scotland Yard complained that fog in London was hampering traffic and preventing the detection of crime.

M

Synoptic
A term used extensively in meteorology to indicate that the data used in preparing a chart (for example) were all obtained at the same time and thus show the state of the atmosphere at a particular moment.

Weather Extremes

Country	Temp.	Location	Date
Maximum temperature			
England	32.8°C	Camden Square (London)	22 May 1922
		Horsham (West Sussex)	29 May 1944
		Tunbridge Wells (Kent)	29 May 1944
		Regent's Park (London)	29 May 1944
Northern Ireland	28.0°C	Knockarevan (Co. Fermanagh)	31 May 1997
Scotland	30.9°C	Inverailort (Highland)	25 May 2012
Wales	29.2°C	Towy Castle (Carmarthenshire)	21 May 1989
Minimum temperature			
England	-9.4°C	Lynford (Norfolk)	4 May 1941
			11 May 1941
Northern Ireland	-6.5°C	Moydamlaght (Co. Londonderry)	7 May 1982
Scotland	-7.7°C	Kinbrace (Highland)	5 May 1981
Wales	-6.1°C	Alwen (Conwy)	1 May 1960
		Alwen (Conwy)	3 May 1967
		St Harmon (Powys)	14 May 1984

Country	Pressure	Location	Date
Maximum pressure			
Eire	1043.0 hPA	Sherkin Island (Co. Cork)	12 May 2012
		Valentia Obsy. (Co. Kerry)	
Minimum pressure			
England	968.0 hPa	Sealand (Cheshire)	8 May 1943

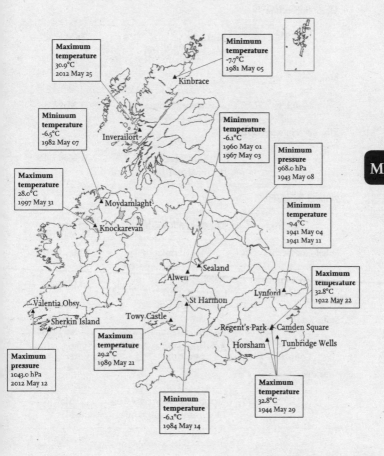

Maximum
temperature
30.9°C
2012 May 25

Minimum
temperature
-7.7°C
1981 May 05

Kinbrace

Minimum
temperature
-6.5°C
1982 May 07

Inverailort

Minimum
temperature
-6.1°C
1960 May 01
1967 May 03

Minimum
pressure
968.0 hPa
1943 May 08

M

Maximum
temperature
28.0°C
1997 May 31

Moydamlaght

Minimum
temperature
-9.4°C
1941 May 04
1941 May 11

Knockarevan

Maximum
temperature
32.8°C
1922 May 22

Sealand

Alwen

Lynford

Valentia Obsy.

St Harmon

Sherkin Island

Towy Castle

Regent's Park   Camden Square

Horsham   Tunbridge Wells

Maximum
pressure
1043.0 hPa
2012 May 12

Maximum
temperature
29.2°C
1989 May 21

Maximum
temperature
32.8°C
1944 May 29

Minimum
temperature
-6.1°C
1984 May 14

The Weather in May 2021

	Observation	Location	Date
Max. temperature			
	25.1°C	Kinlochewe (Ross & Cromarty)	31 May
Min. temperature			
	-6.1°C	St Harmon (Powys)	1 May
Min. overnight temperature			
	12.9°C	Gosport Fleetlands (Hampshire)	9 May
24-hour rainfall			
	103 mm	Mickleden (Cumbria)	21 May
Wind gust			
	81 knots (93 mph)	Needles Old Battery (Isle of Wight)	3 May
Snow depth			
	1 cm	Achiltibuie (Ross & Cromarty)	5 May
Sunshine			
	15.9 hrs	Stornoway (Western Isles)	30 May
		Morecambe (Lancashire)	31 May

Overall, the month of May 2021 was very unsettled, cold and windy. It was also exceptionally wet. Only certain areas of western Scotland had the average rainfall for the month. Elsewhere, it was extremely wet, with well above average rainfall. Some locations, such as eastern Scotland, Wales and south-west England had more than double the normal amount of rain.

The month began exceptionally cold in all areas, with some snow and hail in southern Scotland. There was widespread rain in Wales. (Rainfall in Wales was 245 per cent of the average for May since 1860.) The heavy rain and strong winds in Wales on May 3 and 4 caused road closures, some flooding and power outages, with speed restrictions on various bridges in both north and south Wales. Northern Ireland also experienced showery weather, with some hail and thunderstorms. There were also strong winds in south-western England, causing various transport difficulties in Cornwall. The strong winds also affected the southern counties of England and caused water-supply problems in Hampshire, where some 30,000 properties were without water for several days.

Conditions became slightly warmer in the second week of the month, but thundery showers persisted, with heavy rain and occasionally hail affecting most of Northern Ireland and England at some time. There were very heavy showers in the north of England on May 10 and 11, and some of these crossed the border into southern Scotland. A series of depressions crossed the country, bringing large amounts of rain and high winds to all parts. Heavy rain in the north-east of England on May 16 brought flooding to North Yorkshire, around the area of Settle and Giggleswick. There were exceptionally strong winds in southern England and Wales on May 20 and 21 with disruption to road and rail traffic and bridge closures (including the M48 Severn Bridge). It was also very windy and wet in Northern Ireland, with heavy rain.

The last week of the month saw more settled conditions in all parts of the country, with considerable amounts of sunshine (nearly 16 hours in the Western Isles and in Lancashire). There was persistent fog on the east coast of Scotland on May 29 to 31, only clearing late in the day.

M

Sunrise and Sunset 2022

Location	Date	Rise	Azimuth	Set	Azimuth
Belfast					
	May 01 (Sun)	04:45	62	19:58	298
	May 11 (Wed)	04:25	57	20:16	304
	May 21 (Sat)	04:09	52	20:33	308
	May 31 (Tue)	03:56	48	20:48	312
Cardiff					
	May 01 (Sun)	04:45	64	19:36	296
	May 11 (Wed)	04:27	59	19:52	301
	May 21 (Sat)	04:13	55	20:07	305
	May 31 (Tue)	04:02	52	20:19	308
Edinburgh					
	May 01 (Sun)	04:29	61	19:52	299
	May 11 (Wed)	04:08	55	20:11	305
	May 21 (Sat)	03:50	50	20:30	310
	May 31 (Tue)	03:37	47	20:45	314
London					
	May 01 (Sun)	04:33	64	19:24	296
	May 11 (Wed)	04:15	59	19:41	301
	May 21 (Sat)	04:01	55	19:55	305
	May 31 (Tue)	03:50	52	20:08	308

Note that all times are in Universal Time (UT), otherwise known as Greenwich Mean Time (GMT). These times do not take Summer Time (BST) into account.

Moonrise and Moonset 2022

Location	Date	Rise	Azimuth	Set	Azimuth
Belfast					
	May 01 (Sun)	05:06	62	20:59	303
	May 11 (Wed)	13:56	76	03:20	288
	May 21 (Sat)	01:46	133	09:11	229
	May 31 (Tue)	04:10	43	22:25	320
Cardiff					
	May 01 (Sun)	05:05	64	20:34	301
	May 11 (Wed)	13:49	77	03:02	287
	May 21 (Sat)	01:16	130	09:18	233
	May 31 (Tue)	04:19	47	21:50	315
Edinburgh					
	May 01 (Sun)	04:50	61	20:54	304
	May 11 (Wed)	13:42	76	03:12	289
	May 21 (Sat)	01:44	135	08:51	227
	May 31 (Tue)	03:48	41	22:26	322
London					
	May 01 (Sun)	04:53	64	20:22	301
	May 11 (Wed)	13:37	77	02:50	287
	May 21 (Sat)	01:04	130	09:04	232
	May 31 (Tue)	04:07	47	21:39	315

M

Note that all times are in Universal Time (UT), otherwise known as Greenwich Mean Time (GMT). These times do not take Summer Time (BST) into account.

Twilight Diagrams 2022

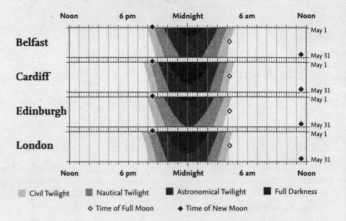

The exact times of the Moon's major phases are shown on the diagrams opposite.

Cyclone

Technically, a name for any circulation of air around a low-pressure centre. (Depressions are also known as 'extratropical cyclones'.) The term is also used specifically for a tropical, revolving storm in the Indian Ocean, known as a 'hurricane' or 'typhoon' in other regions of the world. The term 'tropical cyclones' applies to all such revolving systems.

The Moon's Phases and Ages 2022

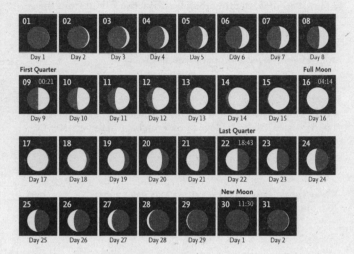

First Quarter — 09 00:21

Full Moon — 16 04:14

Last Quarter — 22 18:43

New Moon — 30 11:30

Day 1	Day 2	Day 3	Day 4	Day 5	Day 6	Day 7	Day 8
Day 9	Day 10	Day 11	Day 12	Day 13	Day 14	Day 15	Day 16
Day 17	Day 18	Day 19	Day 20	Day 21	Day 22	Day 23	Day 24
Day 25	Day 26	Day 27	Day 28	Day 29	Day 1	Day 2

M

Hurricane
The term used for a tropical cyclone in the North Atlantic Ocean or eastern Pacific Ocean. Hurricanes (indeed all tropical cyclones) are driven by high sea-surface temperatures, and cannot occur over the British Isles.

On This Day

3 May 1986 – Heavy rain brought radioactive fallout from the Chernobyl disaster (26 April 1986) to upland areas of Northern Ireland, Wales and Cumbria. The sale of sheep, in particular, from upland areas was banned for several years, because of concerns over the long-lived isotope, caesium-137.

4 May 1955 – The strong winds, with gusts up to 65 mph, create a particularly bad 'Fen blow' in East Anglia, where persistent dry conditions caused the topsoil to dry out and blow away.

7 May 1881 – This was the wettest May day known. No less than 172.2 mm of rain fell on the village of Seathwaite in Cumbria.

9 May 1945 – Following censorship of weather information during the Second World War, this was the first day on which the BBC was able to broadcast information about the weather and give a weather forecast.

9 May 2006 – Cars in counties on the east coast were coated in yellow 'dust', which proved to be pollen from a mass pollination event of birch trees in Denmark that had been carried across the North Sea.

10 May 1818 – A gigantic iceberg, estimated to be nearly 10 km long, became stranded off the coast of Foula, the isolated westernmost of the islands in the Shetlands. Foula itself is about 8 km long.

15 May 1697 – An extremely severe hailstorm (probably the most severe ever recorded in Britain) hits Hitchin and Offley in Hertfordshire. Some hailstones were reported to be up to 110 mm in diameter.

18 May 1891 – There was extreme snowfall on the Whitsun Bank Holiday. A depth of some 15 cm of snow occurred in East Anglia, and the temperature on Ben Nevis that night was recorded as -10°C.

21 May 1950 – The longest track of a British tornado reached from Berkshire to Norfolk. It remained on the ground for 107.1 km, then lifted, becoming a funnel cloud for another 52.6 km.

The Derby Day Disaster, May 1911

The horse race known as the Derby is held at Epsom Downs in Surrey (generally in the latter half of May) but the weather has often been extremely varied, illustrating how fickle the weather may be during that part of the year. On the actual race day there has been torrential rain, blistering heat and exceptionally cold conditions with freezing winds, frost, ice and driving snow.

What was probably the worst disaster occurred on 31 May 1911. That day there were a lot of clear skies and sunshine over southern England, as a result of a generally slack pressure situation, and these conditions continued into the afternoon, proving to be ideal for the creation of severe thunderstorms, several of which were exceptionally strong. The worst of these arrived over the Epsom area of Surrey just as racing finished at about 17:00. The skies darkened and torrential rain began to fall. At nearby Banstead, no more than 3 km away to the east, 62 mm of rain fell in just 50 minutes. (Banstead recorded a total of 91.2 mm of rain over a 24-hour period.) Large hail that was between 37 and 50 mm in diameter fell at Sutton, also in Surrey, about 5 km north-east of the racecourse. There was a phenomenal amount of electrical activity, with no less than 159 lightning strikes counted at Epsom in one 15-minute period between 17:30 and 17:45.

Conditions for people leaving the racetrack were atrocious, with torrential rain and hail that was described as 'the size of walnuts', almost continuous lightning and thunder. The torrential rain not only produced waterlogged ground but also resulted in a sea of mud.

There is some disagreement about the number of people killed by lightning. This confusion has probably arisen because there were casualties from other thunderstorms in southern England that day – although none of the storms was as severe as the one at Epsom. Three people were killed in the immediate vicinity of the racecourse and a large number injured. The lightning set three haystacks on fire, and the torrential rain caused landslips that blocked the railway lines at nearby Merstham and Coulsdon as well as creating severe flooding of the railway at Epsom itself. Overall, the human death toll for southern England that day is most often put at 17, together with four horses (not racehorses) killed by lightning.

M

Front
A zone separating two air masses with different characteristics (typically, with different temperatures and/or humidities). Depressions normally show two fronts: a warm front (where warm air is advancing) and a cold front (where cold air is advancing). The latter normally move faster than warm fronts. When a cold front catches up with a warm front, the warm air is lifted away from the surface and the combined front is known as an occluded front.

June

Introduction

Although meteorologists regard the month of June as marking the start of the summer season, and it may see the longest day at the summer solstice (June 21), in Britain it includes the hottest day just about one year in four. This is about the same as the frequency in the month of August. The hottest day is most common in July. To the general public, June may have come to be associated with the catchphrase 'flaming June', but it is not often accompanied by particularly hot weather, and is only very rarely warmer than July. It shows little sign of getting warmer over time. During the last three centuries, the month has been about as warm as September. In the middle of the month, the weather (when regarded as

showing five seasons), tends to change quite suddenly from 'spring and early summer' to 'high summer'. This sees a quite sudden resumption of predominant westerly conditions from the mixed regimes that prevailed earlier in the year. However, the reduced temperature contrast at this time of year over the Atlantic results in weaker winds and the depressions that arrive from the Atlantic are slow-moving, so any accompanying rain tends to linger and be slow to move away. This striking change in the prevailing weather is often considered to be the start of the 'European monsoon', marked by slow-moving depressions that cross the British Isles and track towards the Baltic or farther northward towards Scandinavia.

High summer – June 18 to September 9

There is a significant change in the overall circulation in early June, from the changeable situation that has prevailed over the preceding three months, when northerly and easterly airflow often predominated. In mid-June, the dominant westerly circulation is restored, with persistent westerly and north-westerly winds and their accompanying depressions. More settled conditions sometimes arise, when the Azores High extends a ridge of high pressure towards western Europe and, occasionally over the British Isles, resulting in a fine, warm, dry summer.

Previous page: Flaming June, *the most famous painting (often regarded as his masterpiece) by Frederic Leighton, painted in 1895.*

Weather Extremes

Country	Temp.	Location	Date
Maximum temperature			
England	35.6°C	Mayflower Park, Southampton (Hampshire)	28 Jun. 1976
Northern Ireland	30.8°C	Knockareven (Co. Fermanagh)	30 Jun. 1976
Scotland	32.2°C	Ochtertyre (Perth & Kinross)	18 Jun. 1893
Wales	33.5°C	Usk (Monmouthshire)	28 Jun. 1976
Minimum temperature			
England	-5.6°C	Santon Downham (Norfolk)	1 Jun. 1962 3 Jun. 1962
Northern Ireland	-2.4°C	Lough Navar Forest (Co. Fermanagh)	4 Jun. 1991
Scotland	-5.6°C	Dalwhinnie (Inverness-shire)	9 Jun. 1955
Wales	-4.0°C	St Harmon (Powys)	8 Jun. 1985

Country	Pressure	Location	Date
Maximum pressure			
Eire	1043.1 hPa	Clones (Co. Monaghan)	14 Jun. 1959
Minimum pressure			
Scotland	968.4 hPa	Lerwick (Shetland)	28 Jun. 1938

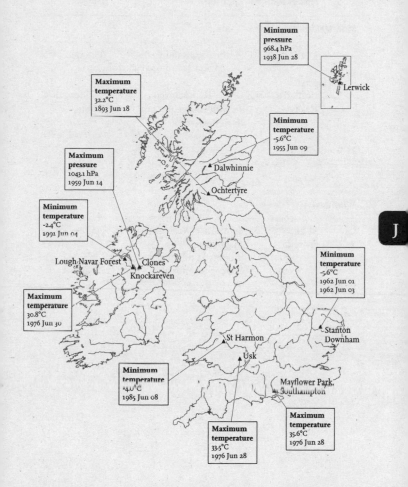

Minimum pressure
968.4 hPa
1938 Jun 28

Maximum temperature
32.2°C
1893 Jun 18

Minimum temperature
-5.6°C
1955 Jun 09

Lerwick

Maximum pressure
1043.1 hPa
1959 Jun 14

Dalwhinnie

Ochtertyre

Minimum temperature
-2.4°C
1991 Jun 04

Lough Navar Forest

Clones

Knockareven

Minimum temperature
-5.6°C
1962 Jun 01
1962 Jun 03

Maximum temperature
30.8°C
1976 Jun 30

Stanton Downham

St Harmon

Usk

Mayflower Park, Southampton

Minimum temperature
-4.0°C
1985 Jun 08

Maximum temperature
33.5°C
1976 Jun 28

Maximum temperature
35.6°C
1976 Jun 28

J°

The Weather in June 2020

Observation	Location	Date
Max. temperature		
33.4°C	Heathrow (London)	25 June
Min. temperature		
-1.9°C	Tulloch Bridge (Inverness-shire)	8 June
24-hour rainfall		
212.8 mm	Honister Pass (Cumbria)	29 June
Wind gust		
56 knots (64 mph)	South Uist (Western Isles) Magilligan (County Londonderry)	20 June
	Warcop (Cumbria)	26 June

Initially, the very sunny and warm weather in late May 2020 continued into June, but conditions soon changed to northerly winds and showery weather. Overall, the month was wet, with more than 144 per cent of average rainfall, especially in western areas, and particularly in Cornwall and Devon, although some south-eastern areas experienced less rain than average, and it was particularly dry in some northern areas of Scotland and much drier than normal in the Shetland Islands.

Showers and longer periods of rain affected north-east Scotland, with some minor flooding on June 5 and 6. Low pressure in depressions brought heavy rain from June 10 to 12, with flooding around Swansea in Wales, with travel problems in Cornwall and Devon on June 11, and difficulties from flooding in north-east England on June 12.

More thundery showers with heavy rain led to flooding in the Midlands on June 15, Major problems, including lightning damage, occurred across a wide area in England and in parts of southern Scotland and Northern Ireland during June 16 to 18, with serious flooding at Pentre in the Rhonda Valley of Wales from a torrential thunderstorm on June 17.

After a sunny period over most of the country, towards the end of the month there was a general breakdown again into showery weather. There was severe weather from June 25 to 27, with heavy rain and thunderstorms. Rail travel in Scotland was disrupted by a landslip, and road travel in the south-west of England was affected by fallen trees. On June 28 it was mainly fine and sunny in the south of England, but there was heavy rain in the north, especially west of the Pennines, where 212.8 mm of rain fell at Honister Pass in Cumbria. After a brighter day on June 29, more persistent rain spread across southern counties of England on June 30.

Sunrise and Sunset 2022

Location	Date	Rise	Azimuth	Set	Azimuth
Belfast					
	Jun 01 (Wed)	03:55	48	20:49	312
	Jun 11 (Sat)	03:48	46	20:59	314
	Jun 21 (Tue)	03:47	45	21:04	315
	Jun 30 (Thu)	03:51	46	21:03	314
Cardiff					
	Jun 01 (Wed)	04:02	52	20:20	309
	Jun 11 (Sat)	03:56	50	20:29	310
	Jun 21 (Tue)	03:56	49	20:34	311
	Jun 30 (Thu)	03:59	49	20:33	310
Edinburgh					
	Jun 01 (Wed)	03:35	46	20:47	314
	Jun 11 (Sat)	03:28	44	20:58	316
	Jun 21 (Tue)	03:27	43	21:03	317
	Jun 30 (Thu)	03:31	44	21:02	316
London					
	Jun 01 (Wed)	03:49	51	20:09	309
	Jun 11 (Sat)	03:44	49	20:18	311
	Jun 21 (Tue)	03:43	49	20:23	311
	Jun 30 (Thu)	03:47	49	20:22	311

Note that all times are in Universal Time (UT), otherwise known as Greenwich Mean Time (GMT). These times do not take Summer Time (BST) into account.

Moonrise and Moonset 2022

Location	Date	Rise	Azimuth	Set	Azimuth
Belfast					
	Jun 01 (Wed)	04:44	40	23:22	321
	Jun 11 (Sat)	17:11	116	02:08	251
	Jun 21 (Tue)	00:54	98	12:40	267
	Jun 30 (Thu)	04:25	40	22:40	317
Cardiff					
	Jun 01 (Wed)	04:57	44	22:47	317
	Jun 11 (Sat)	16:49	114	02:04	252
	Jun 21 (Tue)	00:40	97	12:29	267
	Jun 30 (Thu)	04:37	45	22:08	313
Edinburgh					
	Jun 01 (Wed)	04:21	37	23:24	324
	Jun 11 (Sat)	17:04	117	01:54	250
	Jun 21 (Tue)	00:44	98	12:28	267
	Jun 30 (Thu)	04:02	38	22:39	319
London					
	Jun 01 (Wed)	04:44	44	22:35	317
	Jun 11 (Sat)	16:37	114	01:52	252
	Jun 21 (Tue)	00:28	97	12:17	267
	Jun 30 (Thu)	04:24	45	21:56	313

Note that all times are in Universal Time (UT), otherwise known as Greenwich Mean Time (GMT). These times do not take Summer Time (BST) into account.

Twilight Diagrams 2022

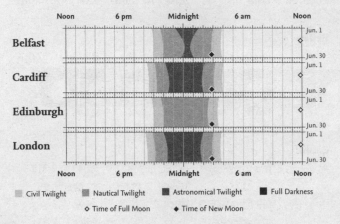

The exact times of the Moon's major phases are shown on the diagrams opposite.

Noctilucent clouds

Clouds seen in midsummer in the middle of the night (the name means 'night-shining') and (for Britain) in the general direction of the North Pole. These clouds (NLC) are seen when the observer is in darkness, but the clouds – which are the highest in the atmosphere at about 185 kilometres, far above all other clouds – are still illuminated by the Sun, which is below the northern horizon. They consist of ice crystals, believed to form around meteoritic dust arriving from space (see page 142).

The Moon's Phases and Ages 2022

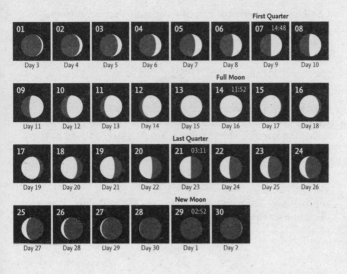

Nacreous clouds
Brilliantly coloured clouds (also known as 'mother-of-pearl' clouds) that are occasionally seen at sunset or sunrise. They occur in the lowest region of the stratosphere at altitudes of 15–30 kilometres. They arise when wave motion at altitude causes water vapour to freeze onto suitable nuclei at very low temperatures (below -83°C).

On This Day

5 June 1983 – A series of extreme tornadic thunderstorms – probably supercell storms that had crossed the English Channel – caused hail, pieces of coke, stones and a crab to fall from the sky on southern counties of England.

8 June 1783 – The eruption of the Laki chain of volcanoes in Iceland began. The results of the eruption caused crop failures across the northern hemisphere. The 'dry haze' caused by the eruption reached Britain on 23 June 1783.

11 June 1963 – Exceptionally heavy rain fell in Dublin, Eire, with 184 mm over the day, 80 mm in one hour, between 15:00 and 16:00.

13 June 1903 – The longest period of continuous rain ever to occur in Britain started on June 13. It rained for 58.5 hours over Camden Square in north London. Rain fell on five of the six days before and after this event, with widespread flooding in London.

17 June 1815 – Torrential rain overnight flooded the ground around Waterloo, in Belgium, delaying Napoleon's offensive, and giving time for the Prussians to reinforce Wellington's army, thus putting to an end the Napoleonic wars.

18 June 1764 – During a storm, a famous landmark, known as 'Lot's Wife', one of the chalk stacks at the Needles at the western end of the Isle of Wight, collapses. Although the lost stack was the origin of the name 'the Needles', the name still persists.

26 June 1953 – On 26 June 1953, Eskdalemuir in Dumfriesshire experienced torrential rain, producing localised flooding and some damage. In 24 hours, 106.0 mm was recorded, with as much as 80.0 mm in one 30-minute period.

28 June 1917 – There was extreme rainfall, amounting to 250 mm at Bruton, Somerset. The amount was second only to the Martinstown event of 18 July 1955.

28 June 1976 – The highest ever June temperature of 35.6°C was recorded at Southampton, Hampshire.

29 June 1853 – John Appleby, coachman to Sir George Cayley, makes the first ever heavier-than-air flight at Brompton Hall, near Scarborough. Appleby gives notice to leave his employment immediately he lands.

The Eruption of Laki in 1783

The eruption of the Icelandic Eyjafjallajökull volcano in 2010 caused disruption to air traffic over Europe, but much earlier eruptions were far more deadly, although poorly understood at the time. Nowadays, with the whole globe under constant surveillance by satellites and instant communication, it is rather difficult to appreciate how it was nearly impossible to understand what was happening thousands of kilometres away, and its implications for the weather and agriculture. The same problem occurred with the later, dramatic effects of the eruption of Tambora in 1815 (see page 108).

Although the Icelandic eruption in 1783 is generally known as the Laki eruption, it was actually what is known to geologists and volcanologists as a fissure eruption, in this case called Lakagigar, which cuts through the Laki volcano. The eruption, from about 130 individual vents, began on June 8 and lasted until 4 February 1784. The fissure was 27 kilometres long. Along with lava, which destroyed some 20 villages, the eruption emitted various gases over a very long period of time, including a vast amount (estimated at no less than 120 million tonnes) of the poisonous gas, sulphur dioxide, most in the early stages of the eruption. The emission of hydrofluoric acid and sulphur dioxide poisoned the soil over most of Iceland, leading to the death of over 50 per cent of the island's livestock and destroying almost all the crops. The resulting famine is estimated to have killed about 25 per cent of Iceland's population.

When the eruption began, air circulating round an unusual high-pressure zone carried the gases in a 'dry haze' first to Norway, then across Prague, Berlin and Paris, reaching England by 23 June. Although this 'dry haze' was noticed and commented upon by many, including the naturalist Gilbert White and the American Benjamin Franklin (in Europe at the time), its source was unknown. The 'dry fog' was not dispersed by sunlight (as a normal moist fog would be), nor by rainfall.

As with the eruption of Tambora, we now know that the emissions from the volcano spread around the northern hemisphere and were to cause a widespread drop in temperatures leading to crop failures and the resultant famines. The amount of material erupted is estimated to be about six times the quantity erupted by Mount Pinatubo in the Philippines on 15 June 1991, which also caused a worldwide drop in temperature. The following winter, 1783–84, was particularly severe in North America and Europe. In a lecture in 1784, Franklin did consider the potential connection between the 'smokes' emitted by the eruption, the prolonged severe winter and the low temperatures that prevailed. He mistakenly ascribed the effects to the combined effects of another Icelandic volcano, Hekla, and an unnamed neighbouring volcano.

J

The Lakagigar fissure today, where it cuts through the Laki volcano.

Noctilucent Clouds

For about one month on either side of midsummer, it is sometimes possible at night (even in the middle of the night) to see glowing clouds when looking in the direction of the pole. These are very high clouds, known as noctilucent clouds. (The name means 'night-shining'.) From Britain they are seen in in the general direction of the North Pole and tend to be most observed from Scotland, although on occasions there are major displays that may be seen from anywhere in the country.

These clouds (NLC) are seen when the observer is in darkness, but the clouds – which are the highest in the atmosphere at about 185 kilometres, far above all other clouds – are still illuminated by the Sun, which is below the northern horizon. (The 'normal' clouds, such as those described on pages 33–42 are found in the lowest layer of the atmosphere, the troposphere, which, even in the tropics, extends to an altitude of no more than 20 km.)

Noctilucent clouds consist of ice crystals, believed to form around particles of meteoritic dust arriving from space. Although they are known to be formed of ice, the origin of the water that freezes into ice crystals is still unknown and the subject of great debate. For many years it was believed that the water also came 'from outside' and was brought by comets and other bodies. There is another possibility. Although water vapour cannot be carried to such great heights, it is possible that water vapour may be created by the breakdown of methane gas, which can rise freely to such extreme altitudes and then be broken down by radiation from the Sun.

Noctilucent clouds, photographed by Alan Tough from Nairn in Scotland, on 31 May 2020, at 00:28.

July

Introduction

The saying that 'the English summer consists of three fine days and a thunderstorm' has been ascribed to both the kings Charles II and George III. However, it is probably a proverbial piece of weather lore, the origins of which are lost. A succession of hot days and humid air certainly provides the conditions for the formation of giant cumulonimbus clouds and thunderstorms with the accompanying torrential rain or even hail. Such a situation usually ends a heatwave (at least for a day or so) and is very typical of British weather.

July is definitely associated with high summer, and it is usually the hottest month of the year. It frequently includes the hottest day of the year. This occurs about 44 per cent of the time, the remaining percentage of hottest days being more-or-less equally divided between June and August. In 1976 it definitely included the hottest day, when the temperature reached 35.9°C at Cheltenham on 3 July 1976.

1976 was the 'drought year', when, beginning in late June, there were ten weeks of sunshine and practically no rain. (It was actually the second driest summer of the twentieth century, after 1995.) A very few locations were fortunate and had some rain, although this was generally less than half of the usual rainfall for the month of July. A high-pressure 'block' over the UK diverted the normal succession of depressions from the Atlantic, and their accompanying rain, south towards the Mediterranean. The drought only came to an end (in late August) just after the government appointed a Drought Minister.

The prolonged drought was actually the result of a very long period of reduced rainfall. In the preceding year, 1975, both the summer and autumn were dry, the winter 1975–76 was particularly dry and then so was the spring of 1976. (Here we use the three-month 'meteorological' seasons, rather than the five seasons that we discuss elsewhere, page 11.) Certain areas of the country actually experienced months with no rain. Portions of south-west England had no rain at all in July and in the first half of August 1976.

July 2006 was the warmest calendar month ever recorded in the Central England Temperature series, which began in 1659. There was an unusual high pressure region over northern Europe and a persistent airstream from the south affecting the British Isles.

Lapse rate
The change in a property with increasing altitude. In meteorology, this is usually the change in temperature. In the troposphere (the lowest layer of the atmosphere), this is a decrease in temperature with an increase in height. This is defined as a positive lapse rate. In the stratosphere (the next higher layer) there is an overall increase in temperature with height, giving a negative lapse rate.

J

Weather Extremes

Country	Temp.	Location	Date
Maximum temperature			
England	37.8°C	Heathrow (London)	1 Jul. 2015
Northern Ireland	30.8°C	Shaw's Bridge, Belfast (Co. Antrim)	12 Jul. 1983
Scotland	32.8°C	Dumfries (Dumfries & Galloway)	20 Jul. 1901 2 Jul. 1908
Wales	34.6°C	Gogerddan (Ceredigion)	19 Jul. 2006
Minimum temperature			
England	-1.7°C	Kielder Castle (Northumberland)	17 Jul. 1965
Northern Ireland	-1.1°C	Lislap Forest (Co. Tyrone)	17 Jul. 1971
Scotland	-2.5°C	Lagganlia (Inverness-shire)	15 Jul. 1977
Wales	-2.5°C	St Harmon (Powys)	9 Jul. 1986

Country	Pressure	Location	Date
Maximum pressure			
Scotland	1039.2 hPa	Aboyne (Aberdeenshire)	16 Jul. 1996
Minimum pressure			
Scotland	967.9 hPa	Sule Skerry (Northern Isles)	8 Jul. 1964

Minimum pressure
967.9 hPa
1964 Jul 08

Sule Skerry

Maximum pressure
1039.2 hPa
1996 Jul 16

Minimum temperature
-2.5°C
1977 Jul 15

Aboyne

Lagganlia

Maximum temperature
32.8°C
1901 Jul 20
1908 Jul 02

Maximum temperature
30.8°C
1983 Jul 12

Dumfries

Kielder Castle

Minimum temperature
-1.°C
1965 Jul 17

Minimum temperature
-1.1°C
1971 Jul 17

Lislap Forest

Shaw's Bridge, Belfast

J

Gogerddan

St Harmon

Heathrow

Maximum temperature
34.6°C
2006 Jul 19

Minimum temperature
-2.5°C
1986 Jul 09

Maximum temperature
37.8°C
2020 Jul 31

The Weather in July 2020

Observation	Location	Date
Max. temperature		
37.8°C	Heathrow (London)	31 July
Min. temperature		
-0.6°C	Kinbrace (Sutherland)	8 July
Min. overnight temperature		
2.4°	Shobdon (Herefordshire)	20 July
24-hour rainfall		
101.8 mm	Aberllefenni, Cymerau Farm (Gwynedd)	4 July
Wind gust		
58 knots (67 mph)	Capel Curig (Gwynedd)	5 July
Sunshine		
14.9 hrs	Shoeburyness (Essex)	12 July
	Morecambe (Lancashire)	19 July

Although July is usually considered to be high summer, the weather is often disappointing. This was certainly the case for July 2020, which was generally poor, with little warmth. A succession of depressions moved across the country, and although high pressure had built up over the near Continent, over France and the Iberian peninsula, and this occasionally extended over the English Channel, bringing better, warmer and sunnier conditions to the southern counties of England, the Midlands and north were subject to persistent heavy rain, dark skies and cool conditions. Scotland generally experienced a wet month, with frequent showers and longer periods of rain. Scottish rainfall was above average, especially in the south-west and north-east. Only Shetland experienced sunnier conditions than average. The weather in Northern Irelend was generally dull with long periods of rain.

At this time of year, depressions forming over the Atlantic tend to be shallow and accompanied by weak winds. (June and July experience fewer gales than any other months in the year.) The depressions themselves are slow-moving and tend to linger for days over an area, taking a long time to move across it. The air is generally warm and humid, so rainfall tends to be heavy, although not as long-lasting as in winter. The high humidity is also a factor in the formation of thunderstorms.

Early in the month, heavy rain and high winds caused problems in both the north and the south of the country. There was flooding in Wales and Cumbria, travel disruption in north Yorkshire, in London, and in the southern counties of England. In the middle of the month, ridges of high pressure invaded the southern counties of England, giving drier, sunny weather, but farther north the weather remained unsettled.

There was an incursion of high pressure at the end of the month and both southern England and Wales experienced warm, sunny conditions. In Northern Ireland and Scotland it was warmer, but both experienced thundery showers.

J

Sunrise and Sunset 2022

Location	Date	Rise	Azimuth	Set	Azimuth
Belfast					
	Jul 01 (Fri)	03:52	46	21:03	314
	Jul 11 (Mon)	04:02	48	20:56	312
	Jul 21 (Thu)	04:16	51	20:43	308
	Jul 31 (Sun)	04:33	56	20:27	304
Cardiff					
	Jul 01 (Fri)	04:00	50	20:33	310
	Jul 11 (Mon)	04:09	51	20:27	308
	Jul 21 (Thu)	04:21	55	20:17	305
	Jul 31 (Sun)	04:35	59	20:02	301
Edinburgh					
	Jul 01 (Fri)	03:32	44	21:01	316
	Jul 11 (Mon)	03:43	46	20:53	314
	Jul 21 (Thu)	03:57	50	20:40	310
	Jul 31 (Sun)	04:15	54	20:22	305
London					
	Jul 01 (Fri)	03:47	49	20:22	310
	Jul 11 (Mon)	03:56	51	20:16	308
	Jul 21 (Thu)	04:08	54	20:06	305
	Jul 31 (Sun)	04:23	58	19:51	301

Note that all times are in Universal Time (UT), otherwise known as Greenwich Mean Time (GMT). These times do not take Summer Time (BST) into account.

Moonrise and Moonset 2022

Location	Date	Rise	Azimuth	Set	Azimuth
Belfast					
	Jul 01 (Fri)	05:33	45	23:04	311
	Jul 11 (Mon)	19:19	139	01:07	226
	Jul 21 (Thu)	23:42	62	14:21	295
	Jul 31 (Sun)	07:11	66	21:54	288
Cardiff					
	Jul 01 (Fri)	05:42	49	22:36	307
	Jul 11 (Mon)	18:45	134	01:15	230
	Jul 21 (Thu)	23:42	64	14:00	293
	Jul 31 (Sun)	07:08	68	21:36	287
Edinburgh					
	Jul 01 (Fri)	05:11	43	23:01	312
	Jul 11 (Mon)	19:20	141	00:47	224
	Jul 21 (Thu)	23:26	61	14:14	296
	Jul 31 (Sun)	06:55	65	21:46	289
London					
	Jul 01 (Fri)	05:29	49	22:24	308
	Jul 11 (Mon)	18:33	135	01:02	230
	Jul 21 (Thu)	23:29	64	13:48	293
	Jul 31 (Sun)	06:56	68	21:24	287

J

Note that all times are in Universal Time (UT), otherwise known as Greenwich Mean Time (GMT). These times do not take Summer Time (BST) into account.

Twilight Diagrams 2022

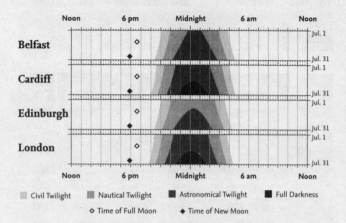

The exact times of the Moon's major phases are shown on the diagrams opposite.

Stratosphere

The second layer in the atmosphere, lying above the troposphere, in which temperatures either stabilise or begin to increase with height. This increase of temperature is primarily driven by the absorption of solar energy by ozone in the ozone layer. In the lowermost region, between the tropopause and about a height of 20 kilometres, the temperature is stable. Above that there is an overall increase to the top of the stratosphere (the stratopause) at an altitude of about 50 kilometres.

The Moon's Phases and Ages 2022

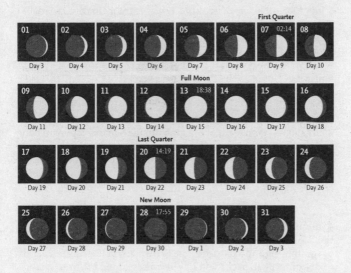

Troposphere
The lowest layer in the atmosphere, in which essentially all weather occurs. It is defined by the way in which temperature declines with height, and is bounded at the top by the tropopause (an inversion at which temperatures either stabilise or begin to increase with height in the overlying stratosphere). The height of the tropopause (and thus the depth of the troposphere) increases from about 7 kilometres at the poles to 14–18 kilometres at the equator.

On This Day

1 July 1968 – The weather this day was extraordinarily varied. Temperatures in the south-east of England reached as high as 33°C, yet there was dense cloud over the Midlands, which produced almost total darkness. Elsewhere, such as in Devon and south Wales , heavy thundery showers were accompanied by large hailstones (some as much as 65 mm across).

2 July 1968 – The stream of hot air from the Sahara responsible for the strange weather the previous day also transported thousands of tonnes of orange-red dust that it deposited over the southern counties of England.

4 July 1925 – The first ever shipping forecast is broadcast by the BBC at 10:30. It soon became a beloved feature of radio programming and a fixture in the British way of life. Any proposals to discontinue it, because it has become redundant, or to change its format or timing, are met with a storm of protest.

8 July 54 BC – An easterly storm wrecked some forty of Julius Caesar's vessels off the Kent coast, thus ending his second attempt to invade Britain. His first attempt, in August the year before, was cancelled after a sudden storm scattered his invasion fleet off Boulogne.

8 July 1746 – Flora MacDonald and Bonnie Prince Charlie (attempting to evade capture, dressed as a lady's maid) sailed in a small boat across the Little Minch from the Western Isles to Skye, encountering a storm on the way. The trip became immortalised in the 'Skye Boat Song'.

15 July – This day is St Swithin's Day. Weather lore states:

> St Swithin's Day: if it does rain
> Full forty days, it will remain
> St Swithin's Day, if it be fair
> For forty days, t'will rain no more

This legend had no basis in actual fact for rain or fair weather. It is perfectly possible to connect a series of 'rain days' associated with various 'rain saints', such that, once begun, it will rain every day of the year.

18 July 1955 – The record for British rainfall during an official observational day (09:00 to 08:59 GMT) was gained by Martinstown, near Dorchester in Dorset, on 18 July 1955, where 279 mm of rain fell in the single day, July 18–19. Most of the rain fell during a 15-hour period.

28 July 2005 – A tornado was recorded in the Birmingham area, being first seen at approximately 14:30 BST. It moved north, and strengthened, causing extensive damage. Its path had an overall length of 12 km. It was the most destructive (and expensive) tornado recorded in Britain.

31 July 2020 – On this day the temperature reached 37.8 °C at Heathrow, Greater London, making this the UK's third hottest day on record. Temperatures across central southern England widely exceeded 34°C and, locally, 35 to 36°C.

J

The Martinstown Rainfall Event

An official meteorological observational day runs from 09:00 one day to 08:59 the next. This applies wherever the observations are made anywhere in the world. Although 09:00 is the standard time for meteorological observations, many stations make observations at other times, such as at 12:00. (Typically, observations are recorded on the hour.) This practice not only enables observations to be compared but ensures that the data for numerical weather forecasting are fully compatible. To this end and to be precise, the times of observations, taken anywhere in the world, are always given in Coordinated Universal Time (UTC), which is identical to Greenwich Mean Time (GMT). Local time, or Summer or Daylight Saving Time is never used.

The record for British rainfall during an official observational day was gained by Martinstown, a village near Dorchester in Dorset, on 18 July 1955, where 279 mm of rain fell in a single observational day (July 18). Practically all of the rain fell during a 15-hour period. Martinstown is more formally known as Winterbourne St Martin. The 'Winterbourne' part of the name indicates that the small stream on which it is located flows mainly during the winter, although unlike many similar streams, this particular stream rarely dries completely during the summer. The underlying rock is chalk, and this acts to absorb rainfall, which is why stream flows are greatest in the winter, when the ground is saturated. In the July 1955 event, much of the rain was absorbed by the ground, but the amount was so great that there was considerable run-off. Martinstown suffered extensive flooding some hours after the peak downpour.

Later analysis by the Meteorological Office suggested that even heavier, unrecorded, rainfalls of over 305 mm probably occurred at Winterbourne Steepleton and Winterbourne Abbas, two villages upstream of Martinstown. The British record for rainfall in any 24 hours (that is, not necessarily beginning and ending at the times defining an official observational day) is the 341.4 mm recorded at Honister Pass in Cumbria on 5 December 2015, delivered by Storm Desmond. This record surpassed the previous total of 317 mm that fell at Seathwaite in Borrowdale, also in Cumbria, on 19–20 November 2009.

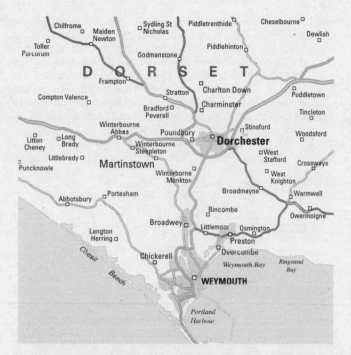

Martinstown (Winterbourne St Martin), the location of the British rainfall record, is near Dorchester in Dorset. Also shown are Winterbourne Stapleton and Winterbourne Abbas where it is believed that the rainfall was even greater.

J

August

Introduction

Although August is always regarded as high summer, it sees the hottest day only as frequently as June. Both have fewer such days than July. August is also a quiet month, with less wind, being similar to July in that respect. However, to many counties in the east of England it is the wettest month as many thunderstorms and their associated rain travel across from the west.

The very first weather forecast appeared in *The Times* on 1 August 1861. It was prepared by Robert Fitzroy – best known to the general public as the captain of HMS *Beagle*, in which Charles Darwin made his momentous voyage. Fitzroy had been appointed in 1854 as 'Statist' to the newly formed Meteorological Department of the Board of Trade. Although, initially, Fitzroy merely collated observations submitted by mariners, he then instigated a system for communicating gale warnings to sailors at various ports. He went on to start 'prognosticating' forthcoming weather, and indeed introduced the term 'weather forecasting'. This resulted in his producing, and *The Times* publishing, the first ever forecast.

This revolutionary forecast was:

General weather probable during next two days:
North – Moderate westerly wind; fine
West – Moderate south-westerly; fine
South – Fresh Westerly; fine

Unfortunately, British weather being so changeable and (in those days) so unpredictable, Fitzroy's forecasts soon became inaccurate and developed into a subject of derision. The situation was not helped by those meteorologists who wanted to put the subject on a scientific basis. Fitzroy's empirical methods were particularly criticised by Francis Galton who, in effect, wished to calculate everything. (Galton is infamous for proposing and promoting eugenics, and famous for producing an equation for obtaining the optimum cup of tea.) As the physics behind meteorology were so poorly understood at that time, Galton believed that true forecasts were impossible. Nevertheless, he remained fascinated by meteorology, and produced the first weather map, discovered anticyclones and proposed a theory explaining their existence.

Fitzroy's storm warning system was discontinued, but reinstated within a short period because both the general public and mariners campaigned for the reintroduction. The weather forecasts were also stopped, and only on 1 April 1875 did *The Times* publish the first weather map, prepared by Galton (see page 177), as distinct from the first weather forecast.

August was originally the sixth month (known as *sextilis*) of the ten-month Roman calendar (which omitted the winter months and began the year in March). It became the eighth month when, in about 700 BC the months January and February were added at the beginning of the year. It originally had 29 days, but gained 2 days when Julius Caesar created the Julian calendar in 46 BC. It was renamed in honour of the emperor Augustus in 8 BC. It retained 31 days when Pope Gregory XIII instigated the calendar reform leading to the Gregorian calendar, introducing the new calendar in 1582.

Until 1965, the August Bank Holiday was held at the beginning (not the end) of the month. The first Monday in August is still a public holiday in Scotland, but in England, Wales and Northern Ireland it has been moved to the last Monday in August.

A

Robert Fitzroy (1805–1865), widely regarded as the founder of modern meteorology. A sea area is named after him, and the headquarters of the Met Office are on Fitzroy Road in Exeter.

Weather Extremes

Country	Temp.	Location	Date
Maximum temperature			
England	38.5°C	Faversham (Kent)	10 Aug. 2003
Northern Ireland	30.6°C	Tandragee Ballylisk (Co. Armagh)	2 Aug. 1995
Scotland	32.9°C	Greycrook (Scottish Borders)	9 Aug. 2005
Wales	35.2°C	Hawarden Bridge (Flintshire)	2 Aug. 1990
Minimum temperature			
England	-2.0°C	Moor House (Cumbria) Kielder Castle (Northumberland)	28 Aug. 1977 14 Aug. 1994
Northern Ireland	-1.9°C	Katesbridge (Co. Down)	24 Aug. 2014
Scotland	-4.5°C	Lagganlia (Inverness-shire)	21 Aug. 1973
Wales	-2.8°C	Alwen (Conwy)	29 Aug. 1959

Country	Pressure	Location	Date
Maximum pressure			
Scotland	1037.4 hPa	Kirkwall (Orkney)	25 Aug. 1968
Minimum pressure			
Eire	967.7 hPa	Belmullet (Co. Mayo)	14 Aug. 1959

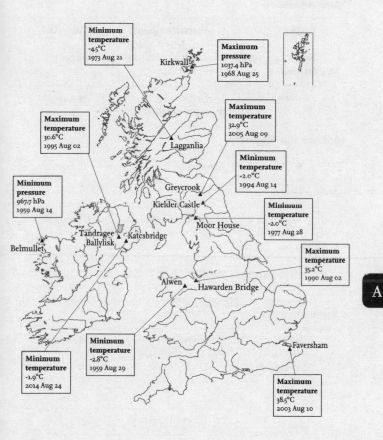

Minimum temperature
-4.5°C
1973 Aug 21

Maximum pressure
1037.4 hPa
1968 Aug 25

Kirkwall

Maximum temperature
32.9°C
2005 Aug 09

Maximum temperature
30.6°C
1995 Aug 02

Lagganlia

Minimum temperature
-2.0°C
1994 Aug 14

Minimum pressure
967.7 hPa
1959 Aug 14

Greycrook

Kielder Castle

Minimum temperature
-2.0°C
1977 Aug 28

Belmullet

Tandragee
Ballylisk
Katesbridge

Moor House

Maximum temperature
35.2°C
1990 Aug 02

Alwen
Hawarden Bridge

A

Faversham

Minimum temperature
-2.8°C
1959 Aug 29

Minimum temperature
-1.9°C
2014 Aug 24

Maximum temperature
38.5°C
2003 Aug 10

The Weather in August 2020

Observation	Location	Date
Max. temperature		
36.4°C	Heathrow (London) Kew Gardens (London)	7 August
Min. temperature		
-0.4°C	Loch Glascarnoch (Ross & Cromarty)	24 August
Min. overnight temperature		
0.1°C	Katesbridge (County Down)	29 August
24-hour rainfall		
103.8 mm	Hollies (Staffordshire)	13 August
Wind gust		
70 knots (81 mph)	Needles (Isle of Wight)	25 August
Sunshine		
14.4 hrs	Edinburgh Fair Isle (Shetland)	8 August

August 2020 was a very unsettled month that was cloudy and wet over most of Britain. It saw some extreme temperatures and heavy downpours with a large number of thunderstorms and hailstorms. There were also some very wet and windy days, especially when two named storms, Ellen and Francis, arrived and governed conditions from August 19–20 for about a week. Rainfall was about 135 per cent of average for most of the country. Only in north-west Scotland was it drier and sunnier than average.

After westerly weather during the first week, there was an almost unprecedented heatwave in the south-east of England that lasted about 7 days as hot air moved in from the near Continent. Temperatures above 30°C were widespread across England with 34°C and 35°C widely recorded. A thundery easterly airflow then followed, but was soon replaced by westerly weather as a succession of deep depressions crossed the country.

The thunderstorms brought flooding to parts of Wales and both rail and road disruption in Scotland, when a landslip caused a train derailment near Stonehaven with three fatalities. A week later, another landslip blocked the railway line between Shrewsbury and Llandrindod Wells. There was widespread flooding in Cumbria, Lancashire and the Midlands as well as in the south-east, including in Hertfordshire and Suffolk. The arrival of Storm Ellen on August 19 brought flooding and disruption to Northern Ireland. There was flooding in Wales and road problems right across England as far as East Anglia.

Storm Francis caused problems after it arrived on 25 August. There was widespread disruption, with major floods in Glasgow and south-west Scotland, but also farther south, including around Newcastle in the north-east of England, and in Pembrokeshire in Wales, where the south of the country was also badly hit. Flooding led to mass evacuations in north Wales. Two days later, winds and rain brought problems to Northern Ireland and farther south and east there was flooding in the London area, causing rail disruptions. There were also rail closures in Devon and Cornwall, and flooding in Plymouth. It was generally quieter in the last few days of the month.

Sunrise and Sunset 2022

Location	Date	Rise	Azimuth	Set	Azimuth
Belfast					
	Aug 01 (Mon)	04:34	56	20:25	303
	Aug 11 (Thu)	04:52	62	20:05	298
	Aug 21 (Sun)	05:10	67	19:42	292
	Aug 31 (Wed)	05:29	74	19:18	286
Cardiff					
	Aug 01 (Mon)	04:36	59	20:01	301
	Aug 11 (Thu)	04:52	64	19:43	296
	Aug 21 (Sun)	05:08	69	19:23	290
	Aug 31 (Wed)	05:24	75	19:01	285
Edinburgh					
	Aug 01 (Mon)	04:17	55	20:20	305
	Aug 11 (Thu)	04:36	61	19:59	299
	Aug 21 (Sun)	04:56	67	19:35	293
	Aug 31 (Wed)	05:15	73	19:10	286
London					
	Aug 01 (Mon)	04:24	59	19:49	301
	Aug 11 (Thu)	04:40	64	19:32	296
	Aug 21 (Sun)	04:56	69	19:12	291
	Aug 31 (Wed)	05:12	75	18:50	285

Note that all times are in Universal Time (UT), otherwise known as Greenwich Mean Time (GMT). These times do not take Summer Time (BST) into account.

Moonrise and Moonset 2022

Location	Date	Rise	Azimuth	Set	Azimuth
Belfast					
	Aug 01 (Mon)	08:28	76	22:03	278
	Aug 11 (Thu)	20:34	128	03:07	226
	Aug 21 (Sun)	23:20	39	17:06	320
	Aug 31 (Wed)	10:15	105	20:40	251
Cardiff					
	Aug 01 (Mon)	08:22	77	21:49	278
	Aug 11 (Thu)	20:07	125	03:15	230
	Aug 21 (Sun)	23:33	44	16:30	316
	Aug 31 (Wed)	09:57	104	20:36	252
Edinburgh					
	Aug 01 (Mon)	08:14	76	21:54	279
	Aug 11 (Thu)	20:30	130	02:45	224
	Aug 21 (Sun)	22:56	37	17:06	323
	Aug 31 (Wed)	10:06	105	20:25	250
London					
	Aug 01 (Mon)	08:09	77	21:37	278
	Aug 11 (Thu)	19:56	125	03:02	229
	Aug 21 (Sun)	23:19	44	16:19	316
	Aug 31 (Wed)	09:45	104	20:24	252

A

Note that all times are in Universal Time (UT), otherwise known as Greenwich Mean Time (GMT). These times do not take Summer Time (BST) into account.

Twilight Diagrams 2022

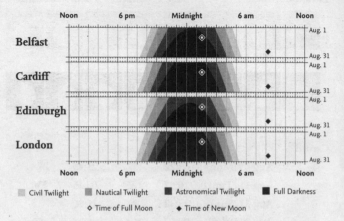

The exact times of the Moon's major phases are shown on the diagrams opposite.

Typhoon

The term used for a tropical cyclone in the western Pacific Ocean. Typhoons are some of the strongest systems encountered anywhere on Earth.

Anticyclone

A high-pressure area. Winds circulate around anticyclones in a clockwise direction in the northern hemisphere. (Anticlockwise in the southern hemisphere.) Anticyclones are slow-moving systems (unlike depressions) and tend to extend their influence slowly from an existing centre.

The Moon's Phases and Ages 2022

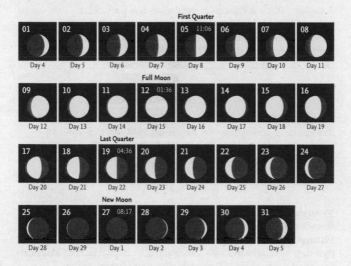

Azores High
A more-or-less permanent high-pressure system in the North Atlantic, generally centred approximately over the islands of the Azores, or closer to Iberia (Portugal and Spain). It arises when air that has risen at the equator descends at the sub-tropical high-pressure zones.

Icelandic Low
A semi-permanent feature of the distribution of pressure over the North Atlantic. Unlike the more-or-less permanent Azores High, it largely arises because depressions (low-pressure systems) frequently pass across the area.

A

On This Day

3 August 2003 – A European heatwave was particularly intense in early August, leading to an estimate of over 70,000 deaths (mainly of elderly people). France was particularly badly hit, with temperatures over most of the country exceeding 39–40°C and nearly 15,000 deaths.

9 August 1843 – An exceptional hailstorm creates a hail swathe, 255 km long across the country, from Stow-on-the-Wold in Gloucestershire to Horsey in Norfolk. In places the hailstones accumulated in piles 1.5 metres high.

14 August 1975 – Known as the Hampstead Storm, an intense thunderstorm created serious flooding in north-west London. Hampstead Climatological Station recorded 170.8 mm of rain that day. Several other observing stations in that part of London recorded more than 100 mm during that observing day.

14 August 1979 – In 1979, the extreme weather turned the Fastnet Race into a disaster. (See page 174.)

15 August 1952 – Torrential rain fell on Exmoor in north Devon. A surge of water, rocks, tree-trunks and other debris fell upon the town of Lynmouth, destroying 28 of the 31 bridges, and more than 100 buildings. A total of 34 people were killed.

16 August 2004 – Fifty-two years after the Lynmouth disaster, a somewhat similar event occurred at Boscastle in Cornwall. The flooding in Boscastle itself was partly the result of the river being dammed by cars, rocks and other debris that blocked the bridge, and this dam suddenly ruptured, releasing a surge of water. The effect was also amplified by the high tide that occurred around 13:00. Some six buildings, 75 cars, five caravans and various boats were washed out to sea, and about 100 buildings destroyed. Unlike at Lynmouth, there was no loss of life, thanks to the response of the rescue services.

19 August 1848 – In one of the worst fishing disasters ever, a storm hit the Moray Firth. It resulted in the loss of over 100 lives of fishermen and 124 boats. The inquiry into the incident led to changes in the design of harbours in the north-east of Scotland and also in methods of boat construction.

24 August 1875 – Matthew Webb became the first to swim across the English Channel without artificial aids. He had to initially delay crossing because of stormy weather, but made it in 22 hrs on his second attempt, although currents meant he couldn't land for 5 hours.

28 August 1588 – Although there had been initial skirmishes in the English Channel, on the night of 28 August the English sent in fireships, causing many of the Spanish ships, anchored near Calais, to cut their anchor cables to escape. A strong southerly gale then drove the Spanish ships up the North Sea. After rounding the north of Scotland, another storm struck the fleet, driving them onto the lee shores of the Hebrides and Ireland. Because of the loss of their anchors, most of the fleet was unable to avoid being wrecked.

The Fastnet Race of 1979

The yacht race known as the Fastnet Race is a major event in the international yacht-racing calendar. It is considered a classic offshore racing event. It is held every two years and in 1979 it was also part of the Admiral's Cup series. The latter was an (unofficial) championship for ocean racing. The original Fastnet Race took place over a course of 608 nautical miles (1126 km). It started from Cowes, in the Isle of Wight, passed down the English Channel, crossed the Celtic Sea, then rounded the Fastnet Rock off the south-west coast of Ireland. It then returned south of the Isles of Scilly and ended at Plymouth. The course has subsequently been changed to finish at Cherbourg in France, increasing the length of the race to over 700 nautical miles (1300 km).

The weather for the race is always challenging, because of the succession of depressions that arrive from the Atlantic during the time that the race lasts, and also their differing wind directions. In 1979 the race started on August 11 with 306 yachts. The weather given by the shipping forecast predicted south-westerly winds of force 4 to 5, perhaps increasing to force 6 or 7 for a time. In the event, a very strong depression formed over the Atlantic on August 11–12. This rapidly deepened on August 13, and also deviated from its original track, turning towards the north-east. By August 14 it was centred over Wexford in southern Ireland, where there were gale-force winds. There were extreme winds on the southern flank of this depression over the sea (and thus over the yachts taking part in the race). The central pressure in the depression dropped to
979 hPa and Meteorological Office observations and calculations suggest that the wind over the sea reached force 10, although some of the competitors believed it reached force 11.
The winds and the extreme mountainous seas that had arisen

took a great toll upon the yachts in the race. Of 306 entries, about 100 were knocked down, and no less than 77 rolled upside-down (known to sailors as 'turtling') at least once during the storm.

The largest rescue effort ever known in peacetime was launched, involving Royal Navy, Irish, United States and Dutch Navy ships, six lifeboats, eight Royal Navy helicopters, seven RAF helicopters, two Irish Air Corps aircraft, RAF Nimrod aircraft, civilian tugs, trawlers and tankers. Around 4000 rescuers were involved. The rescue effort took place between the Fastnet Rock and Lands End, and started on August 14 after the winds had dropped to about force 9. The Nimrod aircraft from Kinloss in Scotland acted as Scene of Search Coordinator, until HMS *Broadsword* was ordered to the scene and took over search coordination. Some 136 yachtsmen were rescued but 15 men and three rescuers lost their lives. Only 86 yachts finished the race, 194 being forced to retire, and 24 yachts being abandoned (with five believed to have sunk).

On land there were severe gales in south-western England and west Wales, with a gust of 75 mph recorded at Milford Haven in Pembrokeshire. There was also damage in affected areas on land where four people were killed during the storm.

A somewhat similar disaster hit the Sydney to Hobart yacht race, one over a distance of 630 nautical miles (1170 km), in 1998, when another storm, which also had near-hurricane-force winds, caused five yachts to sink and six people to lose their lives. Of 115 yachts that started the race, only 44 reached Hobart.

A

Below: *The very first weather forecast, prepared by Robert Fitzroy, and published in The Times of 1 August 1861.*

THE WEATHER.

METEOROLOGICAL REPORTS.

Wednesday, July 31, 8 to 9 a.m.	B.	E.	M.	D.	F.	C.	I.	S.
Nairn	29·54	57	56	W.S.W.	6	9	o.	3
Aberdeen	29·60	59	54	S.S.W.	5	1	b.	3
Leith	29·70	61	55	W.	3	5	c.	2
Berwick	29·69	59	55	W.S.W.	4	4	c.	2
Ardrossan	29·73	57	55	W.	5	4	c.	5
Portrush	29·72	57	54	S.W.	2	2	b.	2
Shields	29·80	59	54	W.S.W.	4	5	o.	3
Galway	29·83	65	62	W.	5	4	c.	4
Scarborough	29·85	59	56	W.	3	6	c.	2
Liverpool	29·91	61	56	S.W.	2	8	c.	2
Valentia	29·37	62	60	S.W.	2	5	o.	3
Queenstown	29·88	61	59	W.	3	5	c.	2
Yarmouth	30·05	61	59	W.	5	2	c.	3
London	30·02	62	56	S.W.	3	2	b.	—
Dover	30·04	70	64	S.W.	3	7	o.	2
Portsmouth	30·01	61	59	W.	3	6	o.	2
Portland	30·03	63	59	S.W.	3	2	c.	3
Plymouth	30·00	62	59	W.	5	1	b.	4
Penzance	30·04	61	60	S.W.	2	6	c.	3
Copenhagen	29·94	64	—	W.S.W.	2	6	c.	3
Helder	29·99	63	—	W.S.W.	6	5	c.	3
Brest	30·09	60	—	S.W.	2	6	c.	5
Bayonne	30·13	63	—	—		9	m.	5
Lisbon	30·18	70	—	N.N.W.	4	3	b.	2

General weather probable during next two days in the—
North—Moderate westerly wind ; fine.
West—Moderate south-westerly ; fine.
South—Fresh westerly ; fine.

Explanation.
B. Barometer, corrected and reduced to 32° at mean sea level ; each 10 feet of vertical rise causing about one-hundredth of an inch diminution, and each 10° above 32° causing nearly three-hundredths increase. E. Exposed thermometer in shade. M. Moistened bulb (for evaporation and dew-point). D. Direction of wind (true—two points left of magnetic). F. Force (1 to 12—estimated). C. Cloud (1 to 9). I. Initials :—b., blue sky ; c., clouds (detached); f., fog ; h., hail ; l., lightning ; m., misty (hazy) ; o., overcast (dull); r., rain ; s., snow ; t., thunder. S. Sea disturbance (1 to 9).

Next page: *The first weather map, prepared by Francis Galton, and published in The Times of 1 April 1875, 14 years after Fitzroy's forecast.*

WEATHER CHART, MARCH 31, 1875.

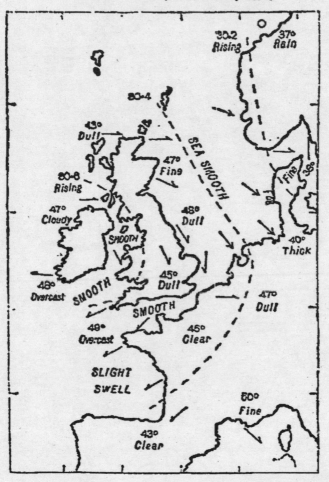

September

Introduction

September has always been regarded as a transitional month, between the high summer of July and August and the autumn. It is a time for late holidays and generally fairly quiet weather. The month tends to be drier than the preceding month, August, and the following one, October. Although there is often a period of windy weather, sometimes with gale-force winds, generally late in the month, the mariners' myth of 'equinoctial gales' certainly does not apply (see the discussion of this point under March, page 80).

To meteorologists, September is the first month of autumn. However, if the year is regarded as consisting of five seasons, the transition to autumn occurs a bit later, after the first week of the month, in which the weather has often been as warm as in June. Because the month includes the equinox (September 23 in 2022), the length of daylight begins to shorten noticeably, and the longer time after sunset allows evening and overnight temperatures to fall. In quiet weather, temperatures soon fall to the saturation point, and dew forms on the ground and vegetation. Temperatures fall even farther overnight, so that there is often some morning mist and fog. It was in mid-September that

---

Autumn – September 10 to November 19
This season is marked by an early period when settled weather may occur, as high pressure makes an occasional incursion across the country from the south or south-west. It is, however, usually marked by a period of wet and windy weather, although this tends to die down as the season transitions in the middle of November into that of the early winter.

John Keats composed his well-loved poem 'To Autumn', with its famous line of 'Season of mists and mellow fruitfulness'.

During the summer, the air arriving from the Atlantic is of the type known to meteorologists as tropical maritime air: cool and fairly humid. The depressions arriving over the country from the Atlantic are fairly shallow, and their winds are subdued. In September, colder air (polar maritime air) begins to come south, and the greater temperature contrast relative to the warmer air from the south tends to deepen depressions and increase their wind speeds.

In Europe, the Full Moon in September is generally known as the 'Harvest Moon'. This is traditionally the Full Moon closest to the equinox. It is also because this is normally the time of the greatest harvest, not only of the various cereal grains, such as wheat and barley, but also apples and similar tree fruits. European harvest festivals were generally held on the Sunday closest to the Full Moon in September.

Sub-tropical highs
Semi-permanent areas in both hemispheres around the latitudes of approximately 30° north and south, where air that has risen at the equator descends back to the surface, becoming heated and dry as it does so. They form the descending limbs of the Hadley cells – the cells closest to the equator (page 226).

S

Weather Extremes

Country	Temp.	Location	Date
Maximum temperature			
England	35.6°C	Bawtry – Hesley Hall (South Yorkshire)	2 Sep. 1906
Northern Ireland	27.8°C	Armagh (Co. Armagh)	1 Sep. 1906
Scotland	32.2°C	Gordon Castle (Moray)	1 Sep. 1906
Wales	31.1°C	Gogerddan (Powys)	1 Sep. 1961
Minimum temperature			
England	-5.6°C	Stanton Downham (Norfolk) Grendon Underwood (Buckinghamshire)	30 Sep. 1969
Northern Ireland	-3.2°C	Magherally (Co. Down)	30 Sep. 1991
Scotland	-6.7°C	Dalwhinnie (Inverness-shire)	26 Sep. 1942
Wales	-5.5°C	St Harmon (Powys)	19 Sep. 1986

Country	Pressure	Location	Date
Maximum pressure			
Northern Ireland	1043 hPa	Ballykelly (Co. Londonderry)	11 Sep. 2009
Minimum pressure			
Eire	957.1 hPa	Claremorris (Co. Mayo)	21 Sep. 1953

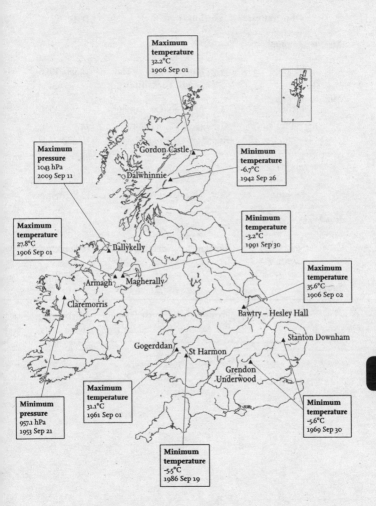

Maximum
temperature
32.2°C
1906 Sep 01

Maximum
pressure
1043 hPa
2009 Sep 11

Minimum
temperature
-6.7°C
1942 Sep 26

Gordon Castle

Dalwhinnie

Minimum
temperature
-3.2°C
1991 Sep 30

Maximum
temperature
27.8°C
1906 Sep 01

Ballykelly

Maximum
temperature
35.6°C
1906 Sep 02

Armagh   Magherally

Claremorris

Bawtry – Hesley Hall

Stanton Downham

Gogerddan   St Harmon

Grendon
Underwood

Minimum
pressure
957.1 hPa
1953 Sep 21

Maximum
temperature
31.1°C
1961 Sep 01

Minimum
temperature
-5.6°C
1969 Sep 30

Minimum
temperature
-5.5°C
1986 Sep 19

S

The Weather in September 2020

Observation	Location	Date
Max. temperature		
31.3°C	Frittenden (Kent)	15 September
Min. temperature		
-5.0°C	Altnaharra (Sutherland)	24 September
	Braemar (Aberdeenshire)	27 September
24-hour rainfall		
88.4 mm	Glen Nevis (Inverness-shire)	13 September
Wind gust		
58 knots	Weybourne (Norfolk)	25 September
(67 mph)	Donna Nook (Lincolnshire)	
Sunshine		
12.5 hrs	Preston Cove House (Dorset)	1 September
	Morecambe (Lancashire)	

For most of the country it was a quiet month. At the very beginning it was particularly wet over Wales, north-west England, Northern Ireland and southern Scotland, with some flooding and travel disruption south-west of Glasgow. In the south-east of England it was generally warm, but there were thundery showers in the north and Midlands.

Most of the country was extremely dry for the whole month, although western Scotland and parts of East Anglia saw higher than average rainfall. It was particularly wet in Norfolk, which experienced high winds and heavy rain towards the end of the month.

During the second week of the month, there was some rain, including heavy showers in the north of England, but the south-east remained fairly warm and dry. In the middle of the month, it turned hot in the south-east, with record temperatures in Kent. North-western Scotland was, however, very wet. Weak depressions from the Atlantic brought changeable weather to the north and west, although the south-east remained largely warm and dry.

High pressure then moved in and gave fine, warm weather for nearly all of Britain, particularly England and Wales, although a nearby depression and its associated fronts affected Northern Ireland and north-western Scotland. The weather in England then turned what might be termed 'autumnal' in the last week of the month. It was generally dull and wet in the north of England. It became very wet and windy on North Sea coasts, particularly in Norfolk, where rainfall was above average. Overnight frosts began to occur in various locations, but the last few days of the month were fairly quiet.

S

Sunrise and Sunset 2022

Location	Date	Rise	Azimuth	Set	Azimuth
Belfast					
	Sep 01 (Thu)	05:31	74	19:16	285
	Sep 11 (Sun)	05:49	81	18:51	279
	Sep 21 (Wed)	06:07	88	18:25	272
	Sep 30 (Fri)	06:24	94	18:02	266
Cardiff					
	Sep 01 (Thu)	05:25	76	18:59	284
	Sep 11 (Sun)	05:41	82	18:37	278
	Sep 21 (Wed)	05:57	88	18:14	272
	Sep 30 (Fri)	06:12	93	17:53	266
Edinburgh					
	Sep 01 (Thu)	05:17	74	19:07	286
	Sep 11 (Sun)	05:36	81	18:41	279
	Sep 21 (Wed)	05:56	88	18:15	272
	Sep 30 (Fri)	06:13	94	17:51	266
London					
	Sep 01 (Thu)	05:13	76	18:48	284
	Sep 11 (Sun)	05:29	82	18:25	278
	Sep 21 (Wed)	05:45	88	18:02	272
	Sep 30 (Fri)	06:00	93	17:41	266

Note that all times are in Universal Time (UT), otherwise known as Greenwich Mean Time (GMT). These times do not take Summer Time (BST) into account.

Moonrise and Moonset 2022

Location	Date	Rise	Azimuth	Set	Azimuth
Belfast					
	Sep 01 (Thu)	11:39	116	20:52	241
	Sep 11 (Sun)	19:33	90	06:49	264
	Sep 21 (Wed)	00:05	44	17:41	311
	Sep 30 (Fri)	12:24	133	19:37	225
Cardiff					
	Sep 01 (Thu)	11:17	114	20:53	243
	Sep 11 (Sun)	19:22	90	06:39	264
	Sep 21 (Wed)	00:14	48	17:13	308
	Sep 30 (Fri)	11:54	129	19:45	229
Edinburgh					
	Sep 01 (Thu)	11:32	117	20:36	240
	Sep 11 (Sun)	19:22	90	06:36	264
	Sep 21 (Wed)	--		17:39	313
	Sep 30 (Fri)	12:22	134	19:16	223
London					
	Sep 01 (Thu)	11:05	114	20:40	243
	Sep 11 (Sun)	19:10	90	06:27	264
	Sep 21 (Wed)	00:01	48	17:01	308
	Sep 30 (Fri)	11:42	129	19:33	229

Note that all times are in Universal Time (UT), otherwise known as Greenwich Mean Time (GMT). These times do not take Summer Time (BST) into account.

S

Twilight Diagrams 2022

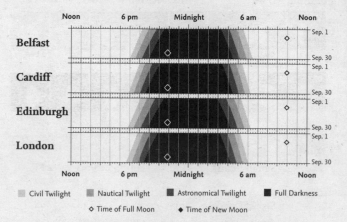

The exact times of the Moon's major phases are shown on the diagrams opposite.

Mesosphere

The third layer of the atmosphere, above the stratosphere and below the thermosphere. It extends from about 50 km (the height of the stratopause) to about 86–100 km (the mesopause). Within it, temperature decreases with increasing altitude, reaching the atmospheric minimum of approximately -123°C at the mesopause. The only clouds occurring within the mesosphere are noctilucent clouds (see page 142).

The Moon's Phases and Ages 2022

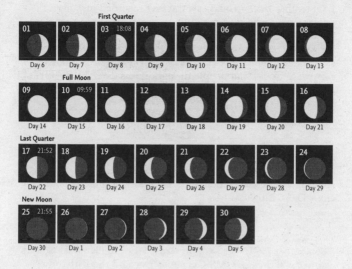

Thermosphere

The fourth layer of the atmosphere, counting from the surface. It is tenuous and lies above the upper limit of the mesosphere, the mesopause, at approximately 86–100 km and extends into interplanetary space. Within it, the temperature increases continuously with height.

S

On This Day

1 September 1859 – Today saw the first recorded instance of a solar flare. Richard Carrington, an amateur astronomer from London, noticed two intensely bright white spots erupting from sunspots. The white flares burned out quickly but this storm caused telegraph failures across the world as well as intense auroras so bright that people thought it was daylight and birds were singing. The solar flare created a geomagnetic storm and has been named the Carrington Event.

2 September 1666 – The Great Fire of London started this Sunday at about 2:00 It lasted until Thursday, September 6. It destroyed about 13,200 houses, which were the homes of some 70,000 of the city's 80,000 inhabitants.

3 September 1691 – A storm off Rame Head in Plymouth Sound claimed 600 lives as HMS *Coronation* ran aground with only 20 survivors. The gunship HMS *Harwich* was also wrecked, killing 300.

5 September 1958 – The heaviest hailstone recorded in England fell at Horsham in Sussex. It was 55 mm in diameter and weighed 190 g.

7 September 1838 – A storm in the North Sea on the Northumberland coast wrecked the steamer *Forfarshire* on the Harcar Rocks off the Farne Islands. This led to 22-year-old Grace Darling and her lighthouse-keeper father taking a small boat and rescuing seven people from the tumultuous seas. Grace Darling became a heroine, feted by Victorian society.

15 September 1819 – Walking in the water meadows at Winchester, the poet John Keats was inspired to write his ode 'To Autumn', regarded by some as his finest work.

17 September 1961 – The remnants of Hurricane Debbie hit Northern Ireland with gusts of over 100 mph. Innumerable trees were blown down, and 13 people killed. The storm (with slightly lower intensity) crossed into Scotland and wreaked destruction from Ayrshire to Caithness.

S

James Glaisher and Balloon Ascents

James Glaisher (1809–1903) was born in Rotherhithe in London, and, early in his life developed an interest in scientific subjects, including astronomy. Initially, he became a surveyor and worked in Ireland. He later said that the Irish clouds that repeatedly impeded his work caused him to become interested in meteorology.

After that initial spell as a surveyor in Ireland, Glaisher turned to astronomical work, moving to Cambridge University Observatory. He moved again to the Royal Observatory Greenwich in December 1835 as assistant to the Astronomer Royal, George Biddell Airy. Glaisher became superintendent of the magnetic and meteorological department, established in 1838. He held this post until he resigned in 1874.

From 1846, Glaisher obtained observations from some 40 stations in Britain, many of which he had established. He was involved with many scientific societies. He was a Fellow of the Royal Astronomical Society and was elected a Fellow of the Royal Society in 1849. In 1850, with other members of those societies, he founded the British Meteorological Society (later the Royal Meteorological Society), and he was its Secretary until 1873, except for 1867–1868 when he was the President. He became a fellow of the Microscopical Society in 1856, its President in 1865 to 1868, and President of the Photographic Society from 1869 to 1892.

Glaisher became famous for his ascents in balloons to obtain meteorological data at altitude, particularly pressure, temperature and humidity. During an ascent, he made observations every 20 seconds. He made numerous ascents, many with Henry Coxwell, the balloonist. One ascent, in particular, brought him notoriety with the general public with both praise and criticism (and even a cartoon in *Punch*). This ascent broke the altitude record. On 5 September 1862, Glaisher ascended from Wolverhampton with Coxwell. At an altitude that he later estimated to be 29,000 feet (approximately 8840 metres – roughly the height of Mount Everest), Glaisher became unconscious (we now know through a lack of oxygen). The balloon rose farther and Glaisher later calculated that it had reached as high as 37,000 feet (11,280 metres). This established a new human altitude record. Coxwell lost the use of his hands,

A photograph of Glaisher (l) and Coxwell (r) in their balloon basket. Glaisher carried numerous meteorological instruments, not clearly seen in this image. © *Rijksuseum*

but was able to pull the cord to the gas-release valve with his teeth, causing the balloon to descend. Neither Glaisher nor Coxwell were permanently injured, although a pigeon that they were carrying did die. The balloon eventually came down on a farm some miles from Ludlow in Shropshire. This episode has since formed the basis of a recent motion picture, *The Aeronauts*, unfortunately romanticised in that Henry Coxwell has been replaced by a fictitious female balloonist.

Glaisher continued to make balloon ascents until 1866. He retired from work at the Royal Observatory in 1874, partly as a result of disagreement with the Director, Airy and also because of his age.

October

Introduction

October is definitely an autumnal month. It generally sees an increase in the number of depressions advancing across the country from the Atlantic. These are often vigorous, with strong winds and carrying plenty of rain. The nature of the accompanying weather depends on the type of air within the depressions. Frequently this will include warm, moist maritime tropical air, and give rise to dull days, often accompanied by extensive rain, because the air has passed over a relatively warm sea and thus taken up significant amounts of moisture. The unstable air behind the cold front of depressions often forms showers.

Because the Arctic is now rapidly cooling, any air arriving from the north may be extremely cold. When there is an incursion of frigid maritime Arctic air, the weather may include the first snow, as well as producing significant heavy showers, which may turn thundery and even turn into hailstorms. Scotland tends to experience more thunderstorms at this time, when strong showers arise in the unstable maritime Arctic air behind the cold fronts of depressions, but the air in front of these depressions arises from a warm sea. It may be heavily laden with moisture and give very heavy and prolonged rain ahead of the warm fronts.

Occasionally during the month there may be a quiet anticyclonic period, giving weather that resembles that in September. Generally the month sees the first extensive frosts, except, perhaps, in the very south of England. In Scotland, the trees may lose their leaves in the strong winds without a major display of colour. In England, the winds are often weaker, being farther from the centres of strong depressions, and so the leaves tend to persist on the trees and may provide a brilliant show of colour.

In Europe and Britain, the Full Moon in October was frequently called the 'Hunter's Moon'. This was the time when people prepared for the coming winter, both by hunting game and by slaughtering livestock. Every three years, however, the first Full Moon after the autumnal equinox actually fell in October. It was then customary to call that particular Full Moon the 'Harvest (rather than Hunter's) Moon'. Among the various tribes in North America, there was a tendency to name the October Full Moon to express the idea that it was the time of leaf-fall. Some typical names were 'Leaf-falling Moon', 'Falling Leaves Moon' or 'Fall Moon'. Some names, such as 'White frost on grass Moon' expressed the idea that significant frosts had arrived.

Edmond Halley (1656–1742) is primarily known to the general public as an astronomer, and because of the famous comet named after him, the return of which he predicted, but did not live to see. (This was the very first such prediction.) However, he also made contributions to many different fields of science.

As far as meteorology is concerned, Halley made various voyages – he was commissioned into the Royal Navy – to investigate terrestrial magnetism, but these led to fundamental understanding of wind patterns. His voyage to St Helena in the South Atlantic (returning in May 1678) was of particular significance. From data he obtained on that voyage, he eventually published in 1686 a paper and a chart of the wind directions (particularly the trade winds and monsoons) around the world. This was of fundamental importance in establishing the details of the circulation of the atmosphere. He also found the important relationship between barometric pressure and height above sea level.

Edmond Halley (1656–1742), after whom the Antarctic research station is named, in an undated painting by Thomas Murray.

O

Weather Extremes

Country	Temp.	Location	Date
Maximum temperature			
England	29.4°C	March (Cambridgeshire)	1 Oct. 1985
Northern Ireland	24.1°C	Strabane (Co. Tyrone)	10 Oct. 1969
Scotland	27.4°C	Tillypronie (Aberdeenshire)	3 Oct. 1908
Wales	28.2°C	Hawarden Bridge (Flintshire)	1 Oct. 2011
Minimum temperature			
England	-10.6°C	Wark (Northumberland)	17 Oct. 1993
Northern Ireland	-7.2°C	Lough Navar Forest (Co. Fermanagh)	18 Oct. 1993
Scotland	-11.7°C	Dalwhinnie (Inverness-shire)	28 Oct. 1948
Wales	-9.0°C	St Harmon (Powys)	29 Oct. 1983

Country	Pressure	Location	Date
Maximum pressure			
Scotland	1045.6 hPa	Dyce (Aberdeenshire)	31 Oct. 1956
Minimum pressure			
Scotland	946.8 hPa	Cawdor Castle (Nairn)	14 Oct. 1891

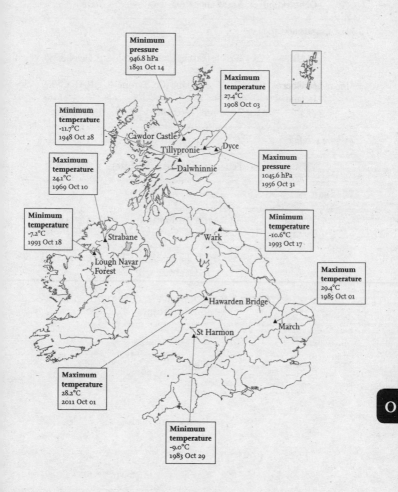

Minimum pressure
946.8 hPa
1891 Oct 14

Maximum temperature
27.4°C
1908 Oct 03

Minimum temperature
-11.7°C
1948 Oct 28

Cawdor Castle

Tillypronie ▲ Dyce

Maximum pressure
1045.6 hPa
1956 Oct 31

Dalwhinnie

Maximum temperature
24.1°C
1969 Oct 10

Minimum temperature
-7.2°C
1993 Oct 18

Strabane

Wark

Minimum temperature
-10.6°C
1993 Oct 17

Lough Navar Forest

Maximum temperature
29.4°C
1985 Oct 01

Hawarden Bridge

St Harmon

March

Maximum temperature
28.2°C
2011 Oct 01

Minimum temperature
-9.0°C
1983 Oct 29

O

The Weather in October 2020

Observation	Location	Date
Max. temperature		
19.1°C	Writtle (Essex)	8 October
Min. temperature		
-3.3°C	Tyndrum (Perthshire)	15 October
24-hour rainfall		
127.1 mm	Fettercairn (Kincardineshire)	4 October
Wind gust		
69 knots (79 mph)	Altnaharra (Sutherland)	25 October

Overall, October 2020 was a very dull and wet month. Parts of southern and eastern England and eastern Scotland received more than double the normal amount of rainfall.

The first of the named storms for the 2020/2021 season, Storm Alex, brought very wet and windy weather to England and Wales, in particular, for the first few days at the very beginning of the month. There was also some flooding and consequent disruption to road and rail traffic in north-east Scotland on the 3rd and 4th of the month with landslips in Fife. Similar flooding affected Northern Ireland, and there were roads blocked by fallen trees in Monmouthshire, with flooding in north Wales. There was also severe flooding in the Midlands in England and in eastern England, with disruption to road and rail traffic. In the London area, people had to be rescued from floods in West Drayton. (Over the month, the London area had more than twice as much rain as normal.) Even in the south-east and south-west of England, more flooding caused difficulties. High winds forced the bridges over the Severn to be closed.

There was more disruption in north-west England a few days later, when heavy showers caused flooding of road and rail lines in Lancashire and Merseyside. Farther south, there were both road and rail problems in mid-Wales.

The next named storm, Barbara, caused more problems after a few quiet days. There was heavy rain and flooding right across Scotland between the 19th and 21st of the month with both road and rail closures. In Northern Ireland it was wet and windy right to the end of the month, with bands of rain and heavy showers crossing the country. Showery weather and bands of rain also affected Wales. The end of the month saw wet and windy weather and heavy, sometimes thundery, showers over the whole of England.

O

Sunrise and Sunset 2022

Location	Date	Rise	Azimuth	Set	Azimuth
Belfast					
	Oct 01 (Sat)	06:26	94	18:00	265
	Oct 11 (Tue)	06:45	101	17:35	259
	Oct 21 (Fri)	07:04	107	17:12	252
	Oct 31 (Mon)	07:24	114	16:50	246
Cardiff					
	Oct 01 (Sat)	06:13	94	17:51	266
	Oct 11 (Tue)	06:30	100	17:28	259
	Oct 21 (Fri)	06:47	106	17:07	253
	Oct 31 (Mon)	07:04	112	16:48	248
Edinburgh					
	Oct 01 (Sat)	06:15	94	17:48	265
	Oct 11 (Tue)	06:36	101	17:22	258
	Oct 21 (Fri)	06:56	108	16:58	252
	Oct 31 (Mon)	07:17	114	16:35	245
London					
	Oct 01 (Sat)	06:02	94	17:39	266
	Oct 11 (Tue)	06:18	100	17:16	259
	Oct 21 (Fri)	06:35	106	16:55	253
	Oct 31 (Mon)	06:53	112	16:36	248

Note that all times are in Universal Time (UT), otherwise known as Greenwich Mean Time (GMT). These times do not take Summer Time (BST) into account.

Moonrise and Moonset 2022

Location	Date	Rise	Azimuth	Set	Azimuth
Belfast					
	Oct 01 (Sat)	13:52	140	20:11	219
	Oct 11 (Tue)	18:11	63	08:36	293
	Oct 21 (Fri)	01:36	65	16:29	289
	Oct 31 (Mon)	14:43	138	21:31	224
Cardiff					
	Oct 01 (Sat)	13:16	135	20:24	224
	Oct 11 (Tue)	18:09	65	08:16	291
	Oct 21 (Fri)	01:34	67	16:11	288
	Oct 31 (Mon)	14:10	134	21:40	228
Edinburgh					
	Oct 01 (Sat)	13:52	142	19:48	217
	Oct 11 (Tue)	17:55	63	08:28	293
	Oct 21 (Fri)	01:20	64	16:22	290
	Oct 31 (Mon)	14:43	140	21:09	222
London					
	Oct 01 (Sat)	13:05	135	20:11	224
	Oct 11 (Tue)	17:57	65	08:03	291
	Oct 21 (Fri)	01:21	67	16:00	288
	Oct 31 (Mon)	13:59	134	21:27	228

O

Note that all times are in Universal Time (UT), otherwise known as Greenwich Mean Time (GMT). These times do not take Summer Time (BST) into account.

Twilight Diagrams 2022

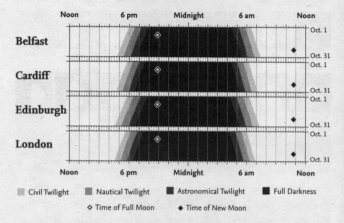

| | Civil Twilight | | Nautical Twilight | | Astronomical Twilight | | Full Darkness |

◇ Time of Full Moon   ◆ Time of New Moon

The exact times of the Moon's major phases are shown on the diagrams opposite.

Aurora

A luminous event occurring in the upper atmosphere between approximately 100 and 1000 km. It arises when energetic particles from the Sun raise atoms to higher energy levels. When the atoms drop back to their original energy level, the emit the characteristic green and red shades (from oxygen and nitrogen, respectively).

The Moon's Phases and Ages 2022

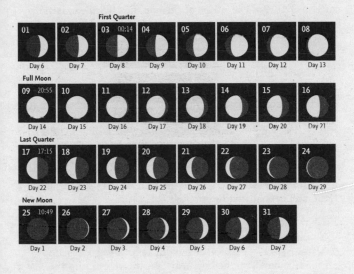

First Quarter

01	02	03 00:14	04	05	06	07	08
Day 6	Day 7	Day 8	Day 9	Day 10	Day 11	Day 12	Day 13

Full Moon

09 20:55	10	11	12	13	14	15	16
Day 14	Day 15	Day 16	Day 17	Day 18	Day 19	Day 20	Day 21

Last Quarter

17 17:15	18	19	20	21	22	23	24
Day 22	Day 23	Day 24	Day 25	Day 26	Day 27	Day 28	Day 29

New Moon

25 10:49	26	27	28	29	30	31
Day 1	Day 2	Day 3	Day 4	Day 5	Day 6	Day 7

Storm surge
A raised level of seawater that is driven ashore and may cause flooding many kilometres inland. The level of the sea is raised primarily by the lower atmospheric pressure at the centre of depressions or tropical cyclones, and may be increased by a high tide (especially a spring tide). The water is often driven ashore by onshore winds. Storm surges have frequently caused many thousands of deaths, especially in vulnerable areas of India and Bangladesh.

O

On This Day

2 October 1697 – The village of Udal on the Hebridean island of North Uist was finally overwhelmed and buried by blowing sand, causing it to be abandoned.

6 October 1822 – A sudden shift in direction of a severe gale over the Isle of Man caused the captain of the Royal Navy vessel *Vigilant* to attempt to put to sea, but the ship was swept onto rocks. Observing the dangerous situation, Sir William Hillary organised two craft to rescue passengers and crew. They saved 97 people. Following another incident ten weeks later, Hillary started to agitate for a national organisation devoted to saving life at sea. This eventually led to the formation of the Royal National Lifeboat Institution (RNLI).

9 October 1799 – HMS *Lutine* sank in a north-westerly gale off the island of Terschelling near Texel in what is now the Netherlands. The ship was carrying £1,200,000 in gold and silver to Hamburg. The loss precipitated the stock market crash that the money was designed to prevent.

14 October 1881 – On 'Black Friday', a storm in the North Sea resulted in the worst Scottish fishing disaster. Nearly 200 fishermen were lost from the east-coast village of Eyemouth alone. The system that created the disaster had also affected Ireland, causing numerous fishing vessels and lives to be lost in those waters.

16 October 1987 – The Great Storm of 1987 occurred on the night of October 16. Extremely high winds caused considerable destruction, with an estimated 15 million trees destroyed, great damage to buildings, and complete disruption to the electricity supply in southern England. Some 19 people were killed, but luckily the worst winds occurred during the night, when most people were in bed. (See pages 210–211.)

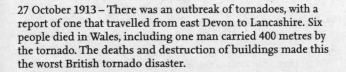

17 October 1091 – The earliest recorded British tornado occurred in the London area. The wooden London Bridge was destroyed, and various properties greatly damaged.

21 October 1966 – Excessive rain over several days mobilised an enormous tip of coal slag above the village of Aberfan in the Taff valley in Wales, causing it to slip. It buried a row of houses and the village school, killing 144 people, 116 of whom were schoolchildren aged 7–10.

25 October 1859 – A violent storm in the Irish Sea wrecked the steam clipper *Royal Charter* on the final leg of its voyage from Melbourne to Liverpool. It was driven ashore on the coast of Anglesey. More than 450 passengers and crew were drowned. The storm caused some deaths onshore and overall about 800 individuals lost their lives. (See pages 208–209.)

O

27 October 1913 – There was an outbreak of tornadoes, with a report of one that travelled from east Devon to Lancashire. Six people died in Wales, including one man carried 400 metres by the tornado. The deaths and destruction of buildings made this the worst British tornado disaster.

The *Royal Charter* Storm

The most extreme storm to occur in the Irish Sea in the nineteenth century was probably the one that happened on 25 and 26 October 1859. A depression, with an extremely low central pressure, approached from the south-west and began to affect the country. It brought extensive damage to Devon and Cornwall, but appears to have changed course towards the north. As it did so, it intensified, affecting the whole of Wales with extreme winds, before eventually crossing Scotland. Winds of Force 12 on the Beaufort scale were experienced, with speeds that are believed to have been well in excess of 100 mph (160 kph). As many as 133 ships were sunk by the storm, and about 90 severely damaged, with many sailors losing their lives.

The worst individual disaster was undoubtedly the loss of the steam clipper *Royal Charter*, which, on the final leg of its voyage from Melbourne to Liverpool, having struggled up the Irish Sea, was suddenly confronted by extreme easterly winds after it had rounded the northern Welsh coast, making for Liverpool. It was driven ashore on the east coast of Anglesey in the early morning of October 26. There were unsubstantiated reports that some returning miners were drowned by being weighed down by the gold in their pockets, and also that some were murdered by landsmen for that self-same gold. More than 450 people were killed in this one incident of the loss of the *Royal Charter*. The storm caused some other deaths onshore and overall about 800 individuals lost their lives as a result of the violent weather.

The loss of the *Royal Charter* had a great effect on the general public, but was of perhaps even greater significance meteorologically, because Captain Robert Fitzroy, then Meteorological Statist to the Board of Trade (a forerunner of the modern Met Office) made an analysis of the storm that he published, most notably, in his book *The Weather Book*, long regarded as the best introduction and textbook for the study of the weather. This episode also induced him to introduce methods of predicting the weather, which he described as 'weather forecasting' – he introduced the term – and also led him to instigate a system to warn sailors of imminent gales and storms. This system consisted of the use of storm cones and drums displayed at various ports to warn mariners of both the direction and severity of forthcoming weather. Although his system was discontinued in the face of criticism of the unscientific nature of his work and predictions, it was soon reinstated following demands from the general public and, in particular, from sailors, especially fishermen, who found it extremely beneficial to have advance warning of possible severe weather when they put to sea.

An imagined image by an unknown painter, of the Royal Charter *clipper ship during the storm in which it was wrecked. Painting now held by the State Library of Queensland.*

O

The Great Storm of 1987

Although the storm that hit southern England on 16 October 1987 is often compared with the Great Storm of 1703 (pages 226–227), in reality, storms of this ferocity are not uncommon over Britain. Normally the centres of such storms pass north of the British Isles, and the extreme winds may affect the north of Scotland. What was unusual about the October 1987 storm was the fact that because of its path, the main impact was felt over south-eastern England. A central pressure of just 957 hPa was recorded at Exeter at 02:00 GMT and an even lower one of 951 hPa over the English Channel. The centre of the depression (which is known technically, and perhaps more dramatically, as an extratropical cyclone) passed from southern Devon, across the counties in the Midlands to the Humber Estuary and then out into the North Sea. The most extreme winds occurred on its southern flank, with gusts of 90 knots (104 mph) at many locations along the South Coast and one recorded gust of 100 knots (115 mph) at Shoreham-by-Sea in West Sussex. (A higher gust of 106 knots (122 mph) has also been reported for Gorleston in Norfolk, although the anemometer there was destroyed by the wind.) The highest hourly mean wind speed recorded was 75 knots (86.5 mph) at the Royal Sovereign Light vessel in the English Channel (south-east of Beachy Head in East Sussex).

It is now known that the extreme winds were produced by a 'sting jet', a phenomenon that was only discovered subsequently, after analysis of the data recorded during the storm. It should be noted that this storm was not 'a hurricane' as widely reported in the media, but that the winds were of 'hurricane force'. (Hurricanes cannot occur over Britain.)

The storm caused widespread damage, and major problems with the electrical grid over southern England, which eventually had to be shut down to prevent a catastrophic failure. The storm was particularly destructive of trees, because many were still in full leaf and the ground was waterlogged and soft from earlier heavy rains. It is estimated that 15 million trees were blown down. Luckily, the worst winds occurred during the night, between 03:00 and 07:00 GMT when few people were outdoors. Although 19 people were killed, the death toll would undoubtedly have been much higher if the extreme winds had struck during the day.

Subsequently, the Meteorological Office was subjected to considerable criticism from the apparent failure to forecast the extreme conditions. In fact, early forecasts, some days beforehand, did suggest the likelihood of extreme winds over southern Britain, but later revisions predicted that the worst of the winds would occur farther south, over France. (In the event, this was true, and the devastation in France was even greater than over England, although fewer individuals were killed.) Later analysis showed that the forecasts were incomplete because of a lack of observations from out in the Atlantic, and there is also the suggestion that industrial action in France prevented data from French stations being passed to the Meteorological Office. The Meteorological Office implemented a number of changes and improvements as a result of their investigations, including alterations to the training of forecasters and modifications to the numerical forecasting models that are used. It also led to the establishment of the National Severe Weather Warning Service. Subsequently, improvements were made to increase the coverage of observations made over the Atlantic, including the deployment of automatic buoys, and increased data acquisition from ships and aircraft.

Trade winds

There are two trade-wind zones, north and south of the equator, with the north-east trades and the south-east trades, respectively, where the air converges on the low-pressure region at the equator. The direction and strength of these winds do remain relatively constant throughout the year, and were thus a reliable source of motive power for sailing ships.

O

November

Introduction

The weather in the early part of November tends to resemble that in October, so it is definitely an autumnal month. However, morning mists and fogs are much more frequent and also, become more persistent, with the decreasing power of the Sun that would otherwise 'burn the fogs off' during the day. Such long-lasting mists and fogs are particularly frequent in the English Midlands and arise either though radiation at night or through the advection of relatively warm, saturated air over cold ground. When such conditions have brought saturated air in an airflow from the west, the descent of the air over the Scottish highlands may result in much warmer temperatures in north-eastern Scotland, particularly in the area around Aberdeen.

Because there is often moderately high pressure over the near Continent at this time of the year, it tends to divert depressions from an easterly track. They then generally veer towards the north, running up the west coast of Britain. The result is heavy rain and strong winds in the west of the country and particularly heavy rainfall (orographic rain) on the mountains of Wales and Scotland. The south and east of England is often much drier

The Full Moon that fell in November, unlike the September 'Harvest Moon' and the October 'Hunter's Moon', did not, in European tradition, have a well-known, universally applied name. It was sometimes known as the 'Frosty Moon' in recognition of the fact that the weather was getting colder, moving towards full winter. On a few occasions we find it called the 'Oak Moon', although this title is more properly applied to the Full Moon in December. Sometimes, if the Full Moon in November was the very last before the winter solstice in December, it was known as the 'Mourning Moon'.

Early winter – November 20 to January 19
This season typically sees an alternation between long periods of mild westerly weather and drier anticyclonic conditions. Roughly half of the years see this pattern, usually with the westerly episodes being wet and windy, with a succession of depressions arriving from the Atlantic. Short cold periods, generally lasting less than a week, occur in between the westerly episodes. Very cold conditions rarely arrive before the very end of January and generally become established in early February.

Weather Extremes

Country	Temp.	Location	Date
Maximum temperature			
England	21.1°C	Chelmsford (Essex) Clacton (Essex) Cambridge (Cambridgeshire) Mildenhall (Suffolk)	5 Nov. 1938
Northern Ireland	18.5°	Murlough (Co. Down)	3 Nov. 1979 1 Nov. 2007 10 Nov. 2015
Scotland	20.6°C	Edinburgh Royal Botanic Garden Liberton (Edinburgh)	4 Nov. 1946
Wales	22.4°C	Trawsgoed (Ceredigion)	1 Nov. 2015
Minimum temperature			
England	-15.5°C	Wycliffe Hall (North Yorkshire)	24 Nov. 1993
Northern Ireland	-12.2°C	Lisburn (Co. Antrim)	15 Nov. 1919
Scotland	-23.3°C	Braemar (Aberdeenshire)	14 Nov. 1919
Wales	-18.0°C	Llysdinam (Powys)	28 Nov. 2010

Country	Pressure	Location	Date
Maximum pressure			
Scotland	1046.7 hPa	Aviemore (Inverness-shire)	10 Nov. 1999
Minimum pressure			
Scotland	939.7 hPa	Monach Lighthouse (Outer Hebrides)	11 Nov. 1877

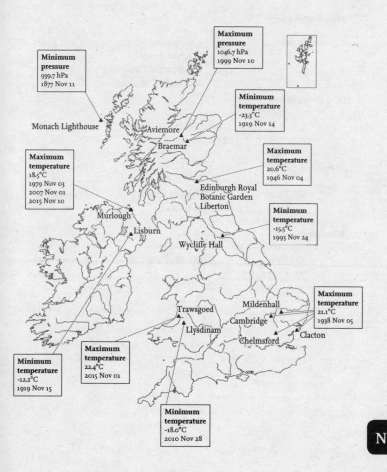

Maximum pressure
1046.7 hPa
1999 Nov 10

Minimum pressure
939.7 hPa
1877 Nov 11

Monach Lighthouse

Minimum temperature
-23.3°C
1919 Nov 14

Aviemore

Braemar

Maximum temperature
18.5°C
1979 Nov 03
2007 Nov 01
2015 Nov 10

Maximum temperature
20.6°C
1946 Nov 04

Edinburgh Royal
Botanic Garden
Liberton

Murlough

Lisburn

Minimum temperature
-15.5°C
1993 Nov 24

Wycliffe Hall

Maximum temperature
21.1°C
1938 Nov 05

Mildenhall

Cambridge

Clacton

Chelmsford

Trawsgoed

Llysdinam

Minimum temperature
-12.2°C
1919 Nov 15

Maximum temperature
22.4°C
2015 Nov 01

Minimum temperature
-18.0°C
2010 Nov 28

N

The Weather in November 2020

Observation	Location	Date
Max. temperature		
18.4°C	Thornes Park (West Yorkshire) Hawarden (Clwyd)	1 November
Min. temperature		
-6.1°C	Aboyne (Aberdeenshire) Cromdale (Morayshire)	29 November
24-hour rainfall		
129.2 mm	Skye Alltdearg House (Inverness-shire)	12 November
Wind gust		
80 knots (92 mph)	Needles (Isle of Wight)	15 November

The month opened with heavy rain, flooding and some loss of power in Scotland, but North Wales was particularly badly hit with flooding and both road and rail disruptions, which also affected Cumbria and northern England. Particularly strong winds were experienced in the south-west of England. The night of November 1 to 2 was extremely mild, but colder weather soon arrived. A week or so later there was more flooding in south-east Wales, and this was followed by wet weather in Northern Ireland with some flooding, and then by flooding in Scotland and, in the third week, some rail disruption in north-east Scotland.

After the initial mild start there was a fairly long period of over a week with southerly and south-westerly winds, when it was mild, but wet and windy. Overall, the month was slightly drier than average in the north-east, but wetter in the north-west, where it was generally dull, as it was over the Home Counties in the south-east of England. It was misty and dull in the Midlands in mid-month, although this then cleared to better conditions although with some showers. At this time it was both wet and windy in Wales.

In Scotland the month saw periods of unsettled weather, with plentiful rain in the south and west, although the north-east was fairly dry. The unsettled weather was interrupted by intervals of quieter, colder conditions and the month ended with much colder and frosty weather early in the day, although some bands of rain spread from the west, mixed with showers. Somewhat similar conditions prevailed in Wales and Northern Ireland, where after a wet and windy period, the end of the month was cooler and calmer, with some frosts, but also some light rain.

N

Sunrise and Sunset 2022

Location	Date	Rise	Azimuth	Set	Azimuth
Belfast					
	Nov 01 (Tue)	07:26	114	16:48	246
	Nov 11 (Fri)	07:46	120	16:29	240
	Nov 21 (Mon)	08:05	125	16:13	235
	Nov 30 (Wed)	08:21	128	16:04	232
Cardiff					
	Nov 01 (Tue)	07:06	112	16:46	247
	Nov 11 (Fri)	07:24	118	16:29	242
	Nov 21 (Mon)	07:41	122	16:16	238
	Nov 30 (Wed)	07:54	125	16:08	235
Edinburgh					
	Nov 01 (Tue)	07:19	115	16:33	245
	Nov 11 (Fri)	07:40	121	16:12	239
	Nov 21 (Mon)	08:01	126	15:56	234
	Nov 30 (Wed)	08:17	130	15:45	230
London					
	Nov 01 (Tue)	06:55	113	16:34	247
	Nov 11 (Fri)	07:12	118	16:17	242
	Nov 21 (Mon)	07:29	122	16:04	238
	Nov 30 (Wed)	07:43	125	15:56	235

Note that all times are in Universal Time (UT), otherwise known as Greenwich Mean Time (GMT). These times do not take Summer Time (BST) into account.

Moonrise and Moonset 2022

Location	Date	Rise	Azimuth	Set	Azimuth
Belfast					
	Nov 01 (Tue)	15:10	130	23:03	233
	Nov 11 (Fri)	17:43	39	11:31	321
	Nov 21 (Mon)	04:32	103	15:13	253
	Nov 30 (Wed)	13:49	113	23:49	251
Cardiff					
	Nov 01 (Tue)	14:42	127	23:07	236
	Nov 11 (Fri)	17:56	43	10:56	316
	Nov 21 (Mon)	04:15	102	15:08	254
	Nov 30 (Wed)	13:29	112	23:45	253
Edinburgh					
	Nov 01 (Tue)	15:07	132	22:43	231
	Nov 11 (Fri)	17:19	36	11:32	323
	Nov 21 (Mon)	04:22	103	14:59	252
	Nov 30 (Wed)	13:42	114	23:34	251
London					
	Nov 01 (Tue)	14:30	127	22:54	236
	Nov 11 (Fri)	17:43	43	10:44	316
	Nov 21 (Mon)	04:03	102	14:56	254
	Nov 30 (Wed)	13:17	112	23:32	252

N

Note that all times are in Universal Time (UT), otherwise known as Greenwich Mean Time (GMT). These times do not take Summer Time (BST) into account.

Twilight Diagrams 2022

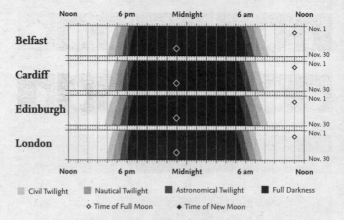

| Civil Twilight | Nautical Twilight | Astronomical Twilight | Full Darkness |

◇ Time of Full Moon   ◆ Time of New Moon

The exact times of the Moon's major phases are shown on the diagrams opposite.

Ionosphere

A region of the atmosphere, consisting of the upper mesosphere and part of the exosphere (from about 60–70 km to 1000 km or more) where radiation from the Sun ionises atoms and causes high electrical conductivity. The ionosphere both reflects certain radio waves back towards the surface, and blocks some wavelengths of radiation from space.

The Moon's Phases and Ages 2022

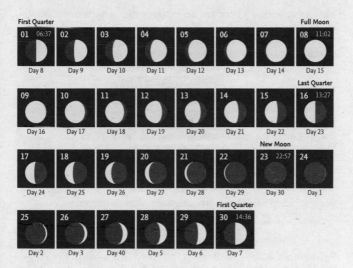

First Quarter

01 06:37	02	03	04	05	06	07	08 11:02
Day 8	Day 9	Day 10	Day 11	Day 12	Day 13	Day 14	Day 15

Full Moon

09	10	11	12	13	14	15	16 13:27
Day 16	Day 17	Day 18	Day 19	Day 20	Day 21	Day 22	Day 23

Last Quarter

17	18	19	20	21	22	23 22:57	24
Day 24	Day 25	Day 26	Day 27	Day 28	Day 29	Day 30	Day 1

New Moon

25	26	27	28	29	30 14:36
Day 2	Day 3	Day 40	Day 5	Day 6	Day 7

First Quarter

Exosphere
The name sometimes applied to the upper region of the thermosphere above an altitude of 200–700 km.

Thermopause
The transitional layer between the underlying mesosphere and the overlying exosphere. It is poorly defined and its altitude lies between 200 and 700 km, depending on solar activity

N

On This Day

November 1665 – There was an extreme drought over the whole country from November 1665 to September 1666. The drought and the heat led to the Great Fire of London (2 September 1666), but the fire and the heat also ended the Great Plague.

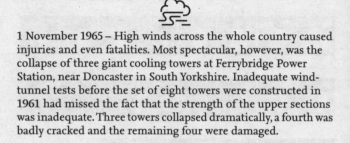

1 November 1965 – High winds across the whole country caused injuries and even fatalities. Most spectacular, however, was the collapse of three giant cooling towers at Ferrybridge Power Station, near Doncaster in South Yorkshire. Inadequate wind-tunnel tests before the set of eight towers were constructed in 1961 had missed the fact that the strength of the upper sections was inadequate. Three towers collapsed dramatically, a fourth was badly cracked and the remaining four were damaged.

9 November 2007 – Predictions indicated that a North Sea surge was likely to occur on November 9 and that the situation would be the worst since 1953 (see pages 76–77). A storm surge of 3 metres above normal levels was predicted. There was some flooding and damage, but no loss of life. At Great Yarmouth, the sea came within just 10 cm of overtopping the sea defences. The Thames Barrier was closed twice that day, before the times of high tide.

18 November 2009 – Between 18 and 20 November 2009, a deep depression and its associated rain brought severe flooding to Cumbria. Seathwaite, in Borrowdale, set a new British record for the amount of rain received in a 24-hour period with 316.4 mm. The rain led to widespread flooding, especially of Cockermouth and Workington.

18 November 1936 – The most extensive, long-lasting fog began to blanket Manchester and persisted until almost the end of the month. It was particularly dense (people in the streets had to navigate by following the tram lines) a week later on November 26.

23 November 1981 – The greatest number of tornadoes (105) were recorded as a cold front moved across England from north Wales to Norfolk. No less than 152 tornadoes were observed in a period of 12 days during this autumn.

26 November 1703 – The greatest British storm in historical times occurred on 26 November 1703 (Old Style) or 7 December 1703 (New Style). It was particularly severe in the southern half of Britain (see pages 226–227).

29 November 1971 – Patchy fog caused a dreadful accident on the M1 motorway. About 50 vehicles were involved, with eight people being killed and some 45 injured.

N

The Great Storm of 1703

The greatest British storm in historical times occurred on 26 November 1703 (Old Style) or 7 December 1703 (New Style). It was particularly severe in the southern half of England and on shipping at sea. The Royal Navy suffered very great losses, including the whole of the Channel Fleet. Merchant shipping was particularly badly hit, and no less than some 700 ships were swept from their moorings in the Pool of London and hurled together in a single mass, blocking the Thames.

There have been various estimates of the overall loss of life, ranging from 8000 to 15,000. The sustained high winds, estimated to have reached about 80 mph (130 kph) caused considerable destruction to standing timber, roofs, chimneys and (in particular) to windmills, over 400 of which were destroyed. Some 2000 chimneys were blown down in London alone, where Queen Anne was forced to take shelter in a cellar at St James' Palace when chimneys and part of the roof collapsed. There was widespread flooding of areas such as the Somerset Levels where some hundreds of people were drowned, with a major loss of livestock. The lowest recorded pressure (in Essex) was 14.1 inches of mercury

---

George Hadley (1685–1768) is often mistaken for his older brother, John, who invented the reflecting quadrant, known to astronomers as the Hadley quadrant. Our George Hadley actually became a barrister, but was primarily interested in the sciences (then known as 'natural philosophy'). He was elected to the Royal Society in 1735, and in May 1735 presented a paper on the nature of the trade winds.

Hadley proposed that heating of the air over the equatorial regions set up an atmospheric circulation, causing air to move towards the equator, but that the rotation of the Earth deflected the motion to produce the north-easterly trades north of the equator and the south-easterly trades in the southern hemisphere. Although his concept of air moving from the poles to the equator in a single circulation is now known to be incorrect – there are three circulation cells in each hemisphere – his name has come to be associated with the cells (the Hadley Cells) on each side of the equator.

(973 hPa), but it has been suggested that the storm's central pressure, over the Midlands, may have been as low as 950 hPa (comparable, or perhaps slightly less than the lowest recorded pressure of 957 hPa for the Great Storm that affected southern England on 16 October 1987, and 953 hPa for the Burns Day Storm of 25–26 January 1990 that was particularly severe over Scotland and northern England).

Accounts of the 1703 disaster are viewed by some as the beginning of modern-style journalism. Daniel Defoe made a particular effort to collect and collate information about the effects of the storm, and his work *The Storm*, published in 1704, is the major source of detailed information about the event. It is also the first comprehensive account of a notable natural event to be published.

A highly imaginative illustration by Robert Chambers (nineteenth century) of the destruction of the Eddystone lighthouse in the great storm of 1703.

N

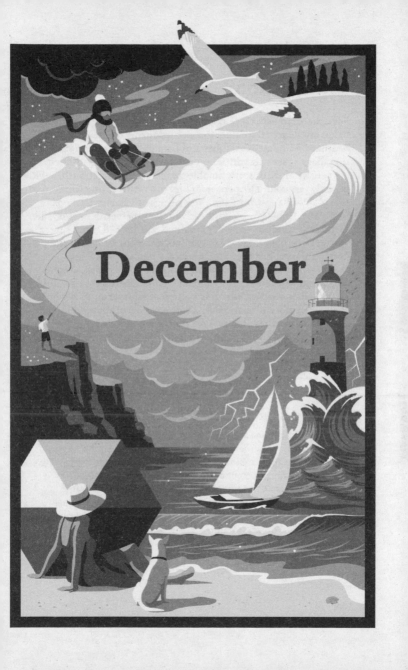

December

Introduction

For meteorologists, December is the beginning of winter. However, based on the actual weather that occurs during the month, it must be linked with late November and January, because, certainly nowadays, it sees very little severe weather. It may – and often is – very windy with some notable storms during the month, but it is rarely very cold. The persistent idea of a white, snowy Christmas is actually a hang-over from the writings of Charles Dickens in particular. In his young days, early in the nineteenth century, the weather around the time of Christmas was indeed more severe. Today, December is rarely as cold as January or February and on average, the month sees just two days when snow is lying.

December may see the winter solstice (December 21 in 2022), and the shortest day, but the temperature remains mild, mitigated by the sea, which is still relatively warm. As a whole then, December is marked by short days, accompanied by wet and windy weather. There have been some notably violent storms that have arrived in December, such as Storm Desmond and Storm Eva in December 2015. The rainfall from Storm

---

Lewis Fry Richardson (1881–1953)
Richardson, a mathematician, was the first to suggest that it would be possible to apply mathematical methods to weather forecasting. He made the first attempt at forecasting forthcoming weather while serving as an ambulance driver in the First World War. Although his initial forecast took far too long, his methods have since been applied to the numerical forecasting used by all principal meteorological organisations.

Desmond broke the United Kingdom's 24-hour rainfall record, with 341.4 mm falling at Honister Pass, Cumbria, on December 5. Rain from Storm Eva added to the problems posed by Storm Desmond, and the precipitation caused severe flooding in Cumbia, where the towns of Appleby, Keswick and Kendal were all flooded on December 22. Some smaller locations were flooded three times during the month.

December is associated with Yule and Yuletide, originally a Germanic festival around the time of the winter solstice. The original pagan festival was taken over by the Christian church, and became Christmas. The Full Moon in December does not have a widely recognised name, unlike the Harvest and Hunter's Moons in September and October, but in the European tradition was sometimes known as 'the Moon before Yule' or even, occasionally, as the 'Wolf Moon', although that name is normally associated with the Full Moon of January.

Gordon Manley (1902–1980)
Gordon Manley's name is well-known to all meteorologists and climatologists. He assembled the Central England Temperature (CET) series of measurements, which is the longest series of monthly mean observations in the world, reaching back to 1659. He also undertook observation of the only British named wind, the Helm wind in Cumbria, working under extremely arduous conditions at the top of Cross Fell (893 metres).

D

Weather Extremes

Country	Temp.	Location	Date
Maximum temperature			
England	17.7°C	Chivenor (Devon)	2 Dec. 1985
		Penkridge (Staffordshire)	11 Dec. 1994
Northern Ireland	16.0°C	Murlough (Co. Down)	11 Dec. 1994
Scotland	18.3°C	Achnashellach (Highland)	2 Dec. 1948
Wales	18.0°C	Aber (Gwynedd)	18 Dec. 1972
Minimum temperature			
England	-25.2°C	Shawbury (Shropshire)	13 Dec. 1981
Northern Ireland	-18.7°C	Castlederg (Co. Tyrone)	24 Dec. 2010
Scotland	-27.2°C	Altnaharra (Highland)	30 Dec. 1995
Wales	-22.7°C	Corwen (Denbighshire)	13 Dec. 1981

Country	Pressure	Location	Date
Maximum pressure			
Scotland	1951.9 hPa	Wick (Caithness)	24 Dec. 1926
Minimum pressure			
Northern Ireland	927.2 hPa	Belfast (Co. Antrim)	8 Dec. 1886

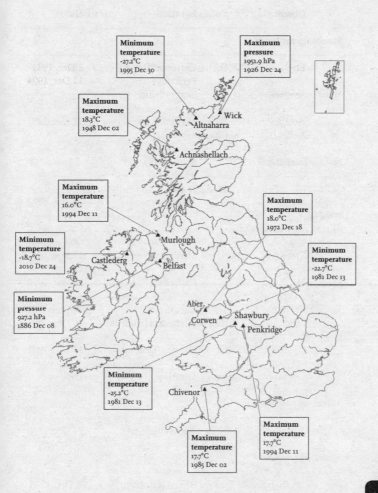

Minimum
temperature
-27.2°C
1995 Dec 30

Maximum
pressure
1951.9 hPa
1926 Dec 24

Maximum
temperature
18.3°C
1948 Dec 02

Wick
Altnaharra

Achnashellach

Maximum
temperature
16.0°C
1994 Dec 11

Maximum
temperature
18.0°C
1972 Dec 18

Minimum
temperature
-18.7°C
2010 Dec 24

Murlough

Castlederg

Belfast

Minimum
temperature
-22.7°C
1981 Dec 13

Minimum
pressure
927.2 hPa
1886 Dec 08

Aber

Corwen

Shawbury

Penkridge

Minimum
temperature
-25.2°C
1981 Dec 13

Chivenor

Maximum
temperature
17.7°C
1985 Dec 02

Maximum
temperature
17.7°C
1994 Dec 11

D

The Weather in December 2020

Observation	Location	Date
Max. temperature		
14.0°C	Prestatyn (Clwyd)	18 December
Min. temperature		
-10.2°C	Dalwhinnie (Inverness-shire)	30 December
24-hour rainfall		
109 mm	Honister Pass (Cumbria)	27 December
Wind gust		
92 knots (109 mph)	Needles (Isle of Wight)	27 December
Snow depth		
18 cm	Loch Glascarnoch (Ross & Cromarty)	31 December

December 2020 was an extremely wet, but not particularly windy, month. Various slow-moving bands of persistent rain brought problems to most parts of the country and many different areas were affected by flooding. Although the month started bright nearly everywhere, with sunny conditions in the south-east of England, it soon turned colder. In the first week of the month there were falls of both sleet and snow over the high ground, and snow was lying at low levels in a few locations. The snow caused some road closures in northern Scotland and even in Yorkshire and the east Midlands. Farther south it was rain that caused disruption. There was flooding in Cambridgeshire, Essex and Suffolk and rail disruption by flooding in East Sussex. Persistent rain caused problems on Merseyside and flooding in Liverpool.

There was more flooding in Scotland on December 11 and then, two days later, in Pembrokeshire in Wales. A few days later it was the turn of south-west England, with the persistent rain causing both road and rail disruptions. Similar areas experienced more flooding yet a few days later. There was some flooding in south-east Wales. The rainfall during the middle of the month was considerable in all parts of the country, exceeding 25 mm nearly everywhere.

From December 23, yet more flooding caused travel disruption in the Midlands. The situation was worsened by the high winds and rain accompanying Storm Bella and there was major disruption in eastern England, especially in north Bedfordshire. The strong winds brought down trees, causing road and rail closures. In Gloucestershire there was widespread flooding and Devon and Cornwall suffered from fallen trees.

Immediately after Christmas, strong wind caused power cuts in northern England, where snow and ice produced difficult conditions. These conditions affected north-west England and then spread south to Yorkshire and the western Midlands, where snow disrupted some public transport.

Northerly winds brought much colder conditions to the whole country at the end of the month. There was widespread snowfall in Scotland, and some sleet and snow in Northern Ireland. There was some snow over high ground in the south-west of England and in the north.

D

Sunrise and Sunset 2022

Location	Date	Rise	Azimuth	Set	Azimuth
Belfast					
	Dec 01 (Thu)	08:22	128	16:03	232
	Dec 11 (Sun)	08:36	131	15:58	229
	Dec 21 (Wed)	08:44	132	15:59	228
	Dec 31 (Sat)	08:46	131	16:07	229
Cardiff					
	Dec 01 (Thu)	07:56	125	16:07	235
	Dec 11 (Sun)	08:08	128	16:04	232
	Dec 21 (Wed)	08:16	128	16:06	232
	Dec 31 (Sat)	08:18	128	16:13	232
Edinburgh					
	Dec 01 (Thu)	08:19	130	15:44	230
	Dec 11 (Sun)	08:33	133	15:39	227
	Dec 21 (Wed)	08:42	134	15:40	226
	Dec 31 (Sat)	08:44	133	15:48	227
London					
	Dec 01 (Thu)	07:45	125	15:55	234
	Dec 11 (Sun)	07:57	128	15:51	232
	Dec 21 (Wed)	08:05	128	15:53	232
	Dec 31 (Sat)	08:07	128	16:01	232

Note that all times are in Universal Time (UT), otherwise known as Greenwich Mean Time (GMT). These times do not take Summer Time (BST) into account.

Moonrise and Moonset 2022

Location	Date	Rise	Azimuth	Set	Azimuth
Belfast					
	Dec 01 (Thu)	13:59	102	next day	
	Dec 11 (Sun)	18:26	43	11:49	319
	Dec 21 (Wed)	06:29	130	14:03	228
	Dec 31 (Sat)	12:35	74	01:47	282
Cardiff					
	Dec 01 (Thu)	13:44	102	next day	
	Dec 11 (Sun)	18:37	47	11:16	314
	Dec 21 (Wed)	06:00	126	14:10	231
	Dec 31 (Sat)	12:30	75	01:31	281
Edinburgh					
	Dec 01 (Thu)	13:50	103	next day	
	Dec 11 (Sun)	18:04	41	11:49	321
	Dec 21 (Wed)	06:26	131	13:43	226
	Dec 31 (Sat)	12:21	73	01:37	282
London					
	Dec 01 (Thu)	13:32	102	next day	
	Dec 11 (Sun)	18:24	47	11:05	315
	Dec 21 (Wed)	05:48	126	13:58	231
	Dec 31 (Sat)	12:18	75	01:19	281

Note that all times are in Universal Time (UT), otherwise known as Greenwich Mean Time (GMT). These times do not take Summer Time (BST) into account.

D

Twilight Diagrams 2022

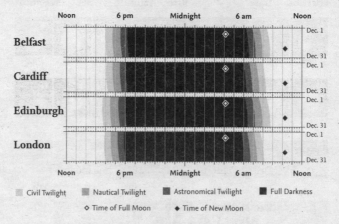

The exact times of the Moon's major phases are shown on the diagrams opposite.

Doldrums

The Doldrums is a zone of reduced winds, generally located over the equatorial region, although moving north and south with the seasons. Air in the Doldrums is largely rising because of solar heating, and horizontal motion across the surface is reduced or non-existent.

The Moon's Phases and Ages 2022

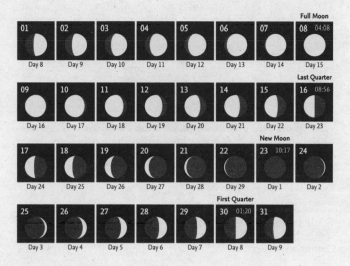

Full Moon

| 01 Day 8 | 02 Day 9 | 03 Day 10 | 04 Day 11 | 05 Day 12 | 06 Day 13 | 07 Day 14 | 08 04:08 Day 15 |

Last Quarter

| 09 Day 16 | 10 Day 17 | 11 Day 18 | 12 Day 19 | 13 Day 20 | 14 Day 21 | 15 Day 22 | 16 08:56 Day 23 |

New Moon

| 17 Day 24 | 18 Day 25 | 19 Day 26 | 20 Day 27 | 21 Day 28 | 22 Day 29 | 23 10:17 Day 1 | 24 Day 2 |

First Quarter

| 25 Day 3 | 26 Day 4 | 27 Day 5 | 28 Day 6 | 29 Day 7 | 30 01:20 Day 8 | 31 Day 9 |

Humidity

A measure of the quantity of water vapour in the air. It generally increases with an increase in temperature.

Adiabatic

Any process in which heat does not enter or leave the system. Air rising in the troposphere generally cools at an adiabatic rate, because it does not lose heat to its surroundings. The fall in temperature is solely because of its expansion: its increase in volume.

D

On This Day

3 December 1933 – In a possibly apocryphal story, Percy Shaw of Halifax, West Yorkshire, was carefully driving home on a perilous road, when he was startled to see the eyes of a cat, shining in the dark. This is credited with the invention of cat's eyes, the reflecting road studs that have been a major contribution to road safety.

4 December 1957 – Dense fog caused the Lewisham rail disaster when two trains collided, wrecking a viaduct on which another train was crossing. The viaduct fell, crushing some of the crashed coaches. Ninety lives were lost, and 173 people injured.

4–5 December 2015 – Storm Denis brought a new British record for the highest 24-hour rainfall – 341.1 mm, at Honister Pass in Cumbria.

5–9 December 1952 – The Great London Smog of 1952 began on December 5 and lasted until December 9. It is believed 4000 people died prematurely as a result.

5 December 2013 – There was a serious surge of the North Sea that was the worst since 1953 (see pages 76–77).

14 December 1820 – The strongest recorded tornado in Britain struck Portsmouth, Hampshire, with a wind speed estimated at between 213 and 240 mph.

18 December 1683 – The most extreme and longest frost ever occurring in Britain began about now. It was so severe that the sea of the English Channel froze the whole way between Dover and Calais. The frost is immortalised in the book *Lorna Doone* by R.D. Blackmore.

21 December 1796 – An easterly gale in Bantry Bay finally destroyed a planned invasion of Ireland by 14,000 French troops. An earlier storm had disrupted plans for the invasion, but the ships eventually made Bantry Bay, only to be dispersed by the unexpected storm.

27 December 1836 – In the only reliably recorded avalanche in Britain, a snow cornice on the chalk clifftop overlooking the town of Lewes in Sussex fell, destroying a row of cottages and killing eight people.

28 December 1879 – The collapse of the railway bridge over the River Tay with the loss of the train crossing at the time resulted in 59 deaths, including everyone on board the train. Only 46 bodies were ever recovered.

D

The Great London Smog of 1952

In early December 1952 a slow-moving anticyclone lay over the British Isles, bringing cold weather and windless conditions. London lay beneath a temperature inversion (the temperature became warmer with height, rather than colder), suppressing the normal turn-over of air. Smoke from the innumerable coal fires, which were being used to keep warm (few had central heating in those days), together with numerous other forms of pollution, such as vehicle exhaust and industrial emissions, was trapped over the city. Particularly injurious to health was the sulphur dioxide emitted into the air. With no wind to disperse the fumes, a dense layer of smog was created over the city. It became particularly dense on December 5, and lasted until December 9. The smog was so dense that it penetrated into buildings, and various theatres and cinemas had to be closed because the audiences were unable to see the stage or screen. Transport was badly affected and only the Underground continued to function. Even walking became almost impossible and street lights failed to penetrate the murk.

At the time, the event was largely tolerated, because London was given to dense fogs, but subsequent analysis suggested that some 4000 people had died prematurely and some additional 100,000 suffered respiratory illness because of the smog. (A much later analysis suggests that the number of excess deaths was much greater, perhaps as many as 12,000.) These findings were primarily responsible for the Clean Air Act, which was eventually passed in 1956. With the introduction of controls on emissions, the introduction of 'smokeless' fuels and the increasing use of gas and electricity for central heating – practically unknown in 1952 – the incidence of such events has decreased, although there was a similar smog 10 years later, in 1962. Since then, however, the infamous London 'pea-soupers' have become a thing of the past. Although dense fogs do still occur, they do not contain such high levels of injurious pollutants.

A photograph of Nelson's Column during the Great London Smog, taken by N.T. Stobbs. Although no time is given, even the street light is only just visible.

D

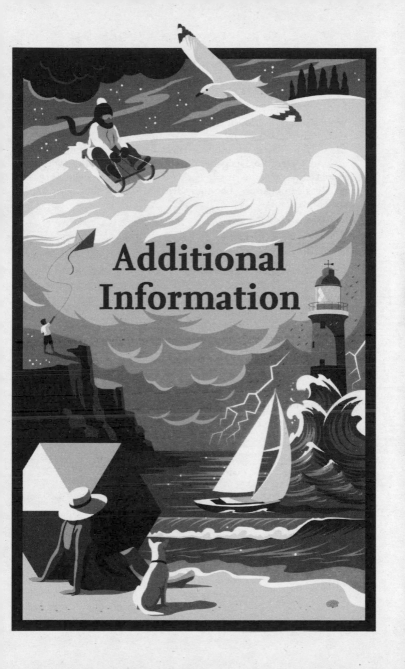

Additional
Information

The Beaufort Scale

Wind strength is commonly given on the Beaufort scale. This was originally defined by Francis Beaufort (later Admiral Beaufort) for use at sea, but was subsequently modified for use on land. Meteorologists generally specify the speed of the wind in metres per second (m s⁻¹). For wind speeds at sea, details are usually given in knots. The equivalents in kph are shown for speeds over land.

The Beaufort scale (for use at sea)

Force	Description	Sea state	Speed	
			Knots	m s⁻¹
0	calm	like a mirror	<1	0.0–0.2
1	light air	ripples, no foam	1–3	0.3–1.5
2	light breeze	small wavelets, smooth crests	4–6	1.6–3.3
3	gentle breeze	large wavelets, some crests break, a few white horses	7–10	3.4–5.4
4	moderate breeze	small waves, frequent white horses	11–16	5.5–7.9
5	fresh breeze	moderate, fairly long waves, many white horses, some spray	17–21	8.0–10.7
6	strong breeze	some large waves, extensive white foaming crests, some spray	22–27	10.8–13.8

The Beaufort scale (for use at sea) – *continued*

Force	Description	Sea state	Speed Knots	m s⁻¹
7	near gale	sea heaping up, streaks of foam blowing in the wind	28–33	13.9–17.1
8	gale	fairly long and high waves, crests breaking into spindrift, foam in prominent streaks	34–40	17.2–20.7
9	strong gale	high waves, dense foam in wind, wave-crests topple and roll over, spray interferes with visibility	41–47	20.8–24.4
10	storm	very high waves with overhanging crests, dense blowing foam, sea appears white, heavy tumbling sea, poor visibility	48–55	24.5–28.4
11	violent storm	exceptionally high waves may hide small ships, sea covered in long, white patches of foam, waves blown into froth, poor visibility	56–63	28.5–32.6
12	hurricane	air filled with foam and spray, visibility extremely bad	≥64	≥32.7

The Beaufort scale (adapted for use on land)

Force	Description	Events on land	Speed km h^{-1}	m s^{-1}
0	calm	smoke rises vertically	<1	0.0–0.21
1	light air	direction of wind shown by smoke but not by wind vane	1–5	0.3–1.5
2	light breeze	wind felt on face, leaves rustle, wind vane turns to wind	6–11	1.6–3.3
3	gentle breeze	leaves and small twigs in motion, wind spreads small flags	12–19	3.4–5.4
4	moderate breeze	wind raises dust and loose paper, small branches move	20–29	5.5–7.9
5	fresh breeze	small leafy trees start to sway, wavelets with crests on inland waters	30–39	8.0–10.7
6	strong breeze	large branches in motion, whistling in telephone wires, difficult to use umbrellas	40–50	10.8–13.8
7	near gale	whole trees in motion, difficult to walk against wind	51–61	13.9–17.1

The Beaufort scale (adapted for use on land) – *continued*

Force	Description	Events on land	Speed km h^{-1}	m s^{-1}
8	gale	twigs break from trees, difficult to walk	62–74	17.2–20.7
9	strong gale	slight structural damage to buildings; chimney pots, tiles, and aerials removed	75–87	20.8–24.4
10	storm	trees uprooted, considerable damage to buildings	88–10	24.5–28.4
11	violent storm	widespread damage to all types of building	102–117	28.5–32.6
12	hurricane	widespread destruction, only specially constructed buildings survive	≥118	≥ 32.7

Twilight Diagrams

Sunrise, sunset, twilight

For each individual month, we give details of sunrise and sunset times for the four capital cities of the various countries that make up the United Kingdom.

During the summer, especially at high latitudes, twilight may persist throughout the night and make it difficult to see the faintest stars. Beyond the Arctic and Antarctic Circles, of course, the Sun does not set for 24 hours at least once during the summer (and rise for 24 hours at least once during the winter). Even when the Sun does dip below the horizon at high latitudes, bright twilight persists throughout the night, so observing the fainter stars is impossible. Even in Britain this applies to northern Scotland, which is why we include a diagram for Lerwick in the Shetland Islands.

As mentioned earlier (page 9) there are three recognised stages of twilight: civil twilight, nautical twilight and astronomical twilight. Full darkness occurs only when the Sun is more than 18° below the horizon. During nautical twilight, only the very brightest stars are visible. During astronomical twilight, the faintest stars visible to the naked eye may be seen directly overhead, but are lost at lower altitudes. They become visible only once it is fully dark. The diagrams show the duration of twilight at the various locations. Of the locations shown, during the summer months there is astronomical twilight for a short time at Belfast, and this lasts longer during the summer at all of the other locations. To illustrate the way in which twilight occurs in the far south of Britain, we include a diagram showing twilight duration at St Mary's in the Scilly Isles. (A similar situation applies to the Channel Islands, which are also in the far south.) Once again, full darkness never occurs.

The diagrams show the times of New and Full Moon (black and white symbols, respectively). As may be seen, at most locations during the year roughly half of New and Full Moon phases may come during daylight. For this reason, the exact phase may be invisible in Britain, but be clearly seen elsewhere in the world. The exact times of the events are given in the diagrams for each individual month.

Belfast, UK – Latitude 54.6°N – Longitude 5.8°W

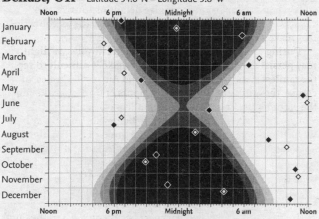

Cardiff, UK – Latitude 51.5°N – Longitude 3.2°W

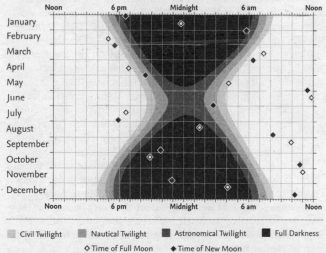

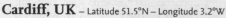

Civil Twilight Nautical Twilight Astronomical Twilight Full Darkness

◇ Time of Full Moon ◆ Time of New Moon

Edinburgh, UK – Latitude 55.9°N – Longitude 3.2°W

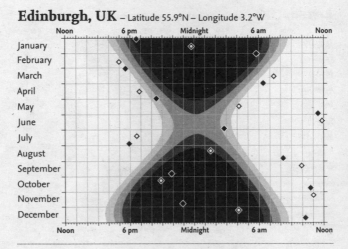

Lerwick, Shetland Islands – Latitude 60.2°N – Longitude 1.1°W

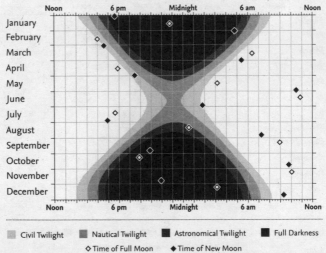

Civil Twilight Nautical Twilight Astronomical Twilight Full Darkness
◇ Time of Full Moon ◆ Time of New Moon

London, UK – Latitude 51.5°N – Longitude 2.0°W

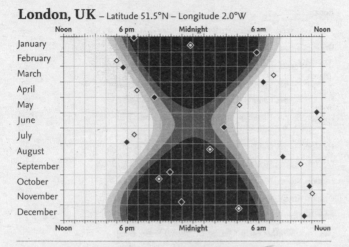

St Mary's, Scilly Isles – Latitude 49.9°N – Longitude 6.4°W

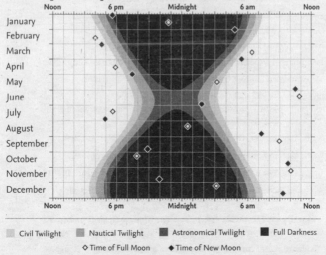

Civil Twilight   Nautical Twilight   Astronomical Twilight   Full Darkness

◇ Time of Full Moon   ◆ Time of New Moon

Further Reading

Books

Chaboud, René, *How Weather Works* (Thames & Hudson, 1996)

Dunlop, Storm, *Clouds* (Haynes, 2018)

Dunlop, Storm, *Collins Gem Weather* (HarperCollins, 1999)

Dunlop, Storm, *Collins Nature Guide Weather* (HarperCollins, 2004)

Dunlop, Storm, *Come Rain or Shine* (Summersdale, 2011)

Dunlop, Storm, *Dictionary of Weather* (2nd edition, Oxford University Press, 2008)

Dunlop, Storm, *Guide to Weather Forecasting* (rev. printing, Philip's, 2013)

Dunlop, Storm, *How to Identify Weather* (HarperCollins, 2002)

Dunlop, Storm, *How to Read the Weather* (Pavilion, 2018)

Dunlop, Storm, *Weather* (Cassell Illustrated, 2006/2007)

Eden, Philip, *Weatherwise* (Macmillan, 1995)

File, Dick, *Weather Facts*, (Oxford University Press, 1996)

Hamblyn, Richard & Meteorological Office, *The Cloud Book: How to Understand the Skies* (David & Charles, 2009)

Hamblyn, Richard & Meteorological Office, *Extraordinary Clouds* (David & Charles, 2009)

Kington, John, *Climate and Weather* (HarperCollins, 2010)

Ludlum, David, *Collins Wildlife Trust Guide Weather* (HarperCollins, 2001)

Meteorological Office, *Cloud Types for Observers* (HMSO, 1982)

Met Office, Factsheets 1–19 (pdfs downloadable from: http://www.metoffice.gov.uk/learning/library/publications/factsheets)

Watts, Alan, *Instant Weather Forecasting* (Adlard Coles Nautical, 2000)

Watts, Alan, *Instant Wind Forecasting* (Adlard Coles Nautical, 2001)

Watts, Alan, *The Weather Handbook* (3rd edn, Adlard Coles Nautical, 2014)

Whitaker, Richard, ed., *Weather: The Ultimate Guide to the Elements* (HarperCollins, 1996)

Williams, Jack, *The AMS Weather Book: The Ultimate Guide to America's Weather* (Univ. Chicago Press, 2009)

Woodward, A., & Penn, R., *The Wrong Kind of Snow* (Hodder & Stoughton, 2007)

Internet links – Current weather

AccuWeather: *http://www.accuweather.com/*
      UK: *http://www.accuweather.com/ukie/index.asp?*

Australian Weather News:
    http://www.australianweathernews.com/

    UK station plots:
      http://www.australianweathernews.com/sitepages/
      charts/611_United_Kingdom.shtml

BBC Weather: *http://www.bbc.co.uk/weather*

CNN Weather: *http://www.cnn.com/WEATHER/index.html*

Intellicast: *http://intellicast.com/*

ITV Weather: *http://www.itv-weather.co.uk/*

Unisys Weather: *http://weather.unisys.com/*

UK Met Office: *http://www.metoffice.gov.uk*

    Forecasts:
      *http://www.metoffice.gov.uk/weather/uk/uk_forecast_*
      weather.html

    Hourly Weather Data:
      http://www.metoffice.gov.uk/education/teachers/
      latest-weather-data-uk

    Latest station plot:
      http://www.metoffice.gov.uk/data/education/chart_latest.gif

    Surface pressure charts:
      http://www.metoffice.gov.uk/public/weather/surface-pressure/

    Explanation of symbols on pressure charts:
      http://www.metoffice.gov.uk/guide/weather/
      symbols#pressure-symbols

    Synoptic & climate stations (interactive map):
      http://www.metoffice.gov.uk/public/weather/climate-network/
      #?tab=climateNetwork

    Weather on the Web:
      http://wow.metoffice.gov.uk/

The Weather Channel:
    http://www.weather.com/twc/homepage.twc

Weather Underground:
    http://www.wunderground.com

Wetterzentrale: *http://www.wetterzentrale.de/pics/Rgbsyn.gif*

Wetter3 (German site with global information):
    http://www.wetter3.de

UK Met Office chart archive:
    http://www.wetter3.de/Archiv/archiv_ukmet.html

General information

Atmospheric Optics:
http://www.atoptics.co.uk/

Hurricane Zone Net:
http://www.hurricanezone.net/

National Climate Data Centre:
http://www.ncdc.noaa.gov/

Extremes:
http://www.ncdc.noaa.gov/oa/climate/severeweather/extremes.html

National Hurricane Center:
http://www.nhc.noaa.gov/

Reading University (Roger Brugge):
http://www.met.reading.ac.uk/~brugge/index.html

UK Weather Information:
http://www.weather.org.uk/

Unisys Hurricane Data:
http://weather.unisys.com/hurricane/atlantic/index.html

World Climate:
http://www.worldclimate.com/

Meteorological Offices, Agencies and Organisations

Environment Canada:
http://www.msc-smc.ec.gc.ca/

European Centre for Medium-Range Weather Forecasting (ECMWF):
http://www.ecmwf.int

European Meteorological Satellite Organisation:
http://www.eumetsat.int/website/home/index.html

Intergovernmental Panel on Climate Change:
http://www.ipcc.ch

National Oceanic and Atmospheric Administration (NOAA):
http://www.noaa.gov/

National Weather Service (NWS):
http://www.nws.noaa.gov/

UK Meteorological Office:
http://www.metoffice.gov.uk

World Meteorological Organisation:
http://www.wmo.int/pages/index_en.html

Satellite images

Eumetsat:
http://www.eumetsat.de/

Image library:
http://www.eumetsat.int/website/home/Images/ImageLibrary/index.html

Group for Earth Observation (GEO):
http://www.geo-web.org.uk/

Societies

American Meteorological Society:
http://www.ametsoc.org/AMS

Australian Meteorological and Oceanographic Society:
http://www.amos.org.au

Canadian Meteorological and Oceanographic Society:
http://www.cmos.ca/

Climatological Observers Link (COL):
https://colweather.ssl-01.com/

European Meteorological Society:
http://www.emetsoc.org/

Irish Meteorological Society:
http://www.irishmetsociety.org

National Weather Association, USA:
http://www.nwas.org/

New Zealand Meteorological Society:
http://www.metsoc.org.nz/

Royal Meteorological Society:
http://www.rmets.org

TORRO: Tornado and Storm Research Organisation:
http://torro.org.uk

Acknowledgements

69	Fridtjof Nansen Archive, National Library of Norway
88	National Gallery, London
93	John Cartwright / Alamy Stock Photo
128	World History Archive / Alamy Stock Photo
141	Wikimedia Commons
143	Alan Tough
163	Liam White / Alamy Stock Photo
176	The Times
177	The Times
193	Rijksmuseum
197	Granger/Shutterstock
209	State Library of Queensland
227	Wikimedia Commons
243	Wikimedia Commons

Index

Notes

Notes

Notes

Notes

Notes

Notes

EXPLORE OUR RANGE OF ASTRONOMY TITLES

Other titles by Storm Dunlop and Wil Tirion

2022 Guide to the Night Sky: Britain and Ireland
978-0-00-839353-3

2022 Guide to the Night Sky: North America
978-0-00-846986-3

2022 Guide to the Night Sky: Southern Hemisphere
978-0-00-846980-1

Latest editions of our bestselling month-by-month guides for exploring the skies. These guides are an easy introduction to astronomy and a useful reference for seasoned stargazers.

Collins Planisphere | 978-0-00-754075-4

Easy-to-use practical tool to help astronomers to identify the constellations and stars every day of the year. For latitude 50°N, suitable for use anywhere in Britain and Ireland, Northern Europe, Canada and Northern USA

Also available

Astronomy Photographer of the Year: Collection 10
978-0-00-846987-0

Winning and shortlisted images from the 2021 Insight Investment Astronomy Photographer of the Year competition, hosted by the Royal Observatory, Greenwich. The images include aurorae, galaxies, our Moon, our Sun, people and space, planets, comets and asteroids, skyscapes, stars and nebulae.

Stargazing | 978-0-00-819627-1

The prefect manual for beginners to astronomy – introducing the world of telescopes, planets, stars, dark skies and celestial maps.

Moongazing | 978-0-00-830500-0

An in-depth guide for all aspiring astronomers and Moon observers, with detailed Moon maps. Covers the history of lunar exploration and the properties of the Moon, its origin and orbit.

The Moon | 978-0-00-828246-2

A celebration of our celestial neighbour, exploring people's fascination with our only natural satellite, illuminating how art and science meet in our profound connection with the Moon.

Northern Lights | 978-0-00-846555-1

Discover the incomparable beauty of the Northern Lights with this accessible guide for both aspiring astronomers and seasoned night sky observers alike.